Björn S

Reading Between the Packets

Björn Scheuermann

Reading Between the Packets

Implicit Feedback in Wireless Multihop Networks

VDM Verlag Dr. Müller

Imprint

Bibliographic information by the German National Library: The German National Library lists this publication at the German National Bibliography; detailed bibliographic information is available on the Internet at http://dnb.d-nb.de.

Cover image: www.purestockx.com

Publisher:
VDM Verlag Dr. Müller Aktiengesellschaft & Co. KG , Dudweiler Landstr. 125 a, 66123 Saarbrücken, Germany,
Phone +49 681 9100-698, Fax +49 681 9100-988,
Email: info@vdm-verlag.de

Zugl.: Düsseldorf, Heinrich-Heine-Universität, Dissertation, 2007

Produced in USA and UK by:
Lightning Source Inc., La Vergne, Tennessee, USA
Lightning Source UK Ltd., Milton Keynes, UK

ISBN: 978-3-8364-5279-3

Abstract

In this thesis, we consider wireless multihop networks with a single channel and omni-directional antennas. Such networks exhibit some distinctive and interesting properties. The most central one is that each transmission is a local broadcast, i. e., it can be received by all nodes in the vicinity. It is thus not exclusive between the sender and the intended receiver. This unique environment turned out to be very challenging for existing communication protocols. Therefore, it has most often been seen—and treated—as a handicap.

We look at the same properties from a very different angle: they can in fact often provide the basis for novel and unconventional solution approaches, and many challenges can be tackled by being aware of their consequences, and by making use of them in tailored protocol designs. The medium properties can be used to obtain information and to co-ordinate actions *implicitly*, i. e., without dedicated information exchange.

As our first major contribution, we introduce a congestion control scheme that bridges what was traditionally transport layer functionality and medium access scheduling. *Implicit hop-by-hop congestion control* does not need to exchange any congestion feedback explicitly. It is based on establishing backpressure with very short queues, by not allowing the transmission of a follow-up packet before further forwarding of the previous one has been overheard. This avoids excessive packet inflow. The concept is realized and assessed in the *Cooperative Cross-layer Congestion Control* (CXCC) protocol.

While CXCC provides congestion control, it does not guarantee TCP-like end-to-end reliability. We thus extend the concept of implicit feedback further, and design the *Backpressure Reliability* (BarRel) transport protocol. BarRel exploits properties of the congestion control approach to infer successful end-to-end packet delivery. In contrast to existing TCP-equivalent transport protocols it does therefore not need a continuous stream of acknowledgments from the destination. Therefore, it largely reduces the amount of control traffic.

Implicit hop-by-hop congestion control can also be extended to a multicast setting. We do so in the *Backpressure Multicast Congestion Control* (BMCC) protocol. Based on implicit feedback and derived from CXCC, it achieves effective source rate adaption at low latencies and minimal control overhead.

Following the discussion of BMCC, we look at network coding, i.e., the combination of multiple packets into one (coded) transmission by intermediate routers. Opportunistic network coding has been proposed to practically increase the capacity of wireless multi-hop networks, but it depends on the spontaneous emergence of situations in which this is possible. Here, we introduce *Near-Optimal Co-ordinated Coding* (noCoCo) for bidirectional wireless multihop data flows. noCoCo demonstrates how—through implicitly co-ordinated scheduling—the existence of coding opportunities can be guaranteed.

Finally, we show that implicitly obtained information can not only be used in the design of protocols, but also to overcome other difficulties. In experiments with network protocols, a common time basis of the nodes' log files is vital for the evaluation of results. Exchanging time information—as time synchronization protocols like NTP do—may, however, interfere with the network traffic generated in the experiment. We thus introduce an alternative in form of a post-facto time synchronization method, which is based on implicitly obtained information. It takes event log files from the participating nodes as its input and uses parallel observations of the same events by multiple nodes to infer the relative deviation of the clocks. This allows to compute globally consistent timestamps, without a need for dedicated communication during the experiment.

Acknowledgments

While this thesis bears only my name on its cover, it has in fact been influenced by, shaped through, and constructed from the contributions of many. This not only applies to the innumerable researchers whose previous work laid the foundations for my own humble contributions, it is also more than true for all the people who supported me during my work.

The first and foremost person to be mentioned here is my doctoral adviser, Martin Mauve. His insights and suggestions, his encouragement, and his tremendous support were invaluable. He has truly been the best mentor one could hope for, in each and every sense.

I want to thank my other two referees, Stefan Conrad and Peter Martini, for taking the time to read this thesis, and for finishing their reviews much sooner than I had ever dared to hope.

I am also deeply indebted to my marvelous colleagues, co-authors, and friends at the Computer Networks and Communication Systems group in Düsseldorf. The countless discussions with my fellow doctoral students, especially with Christian Lochert, Wolfgang Kiess, Michael Stini, Jedrzej Rybicki, and Tran Thi Minh Chau, were a never dwindling source of ideas and inspiration. Our many collaboratively authored publications are designative for the open and creative atmosphere, which I have enjoyed more than anything else—and they also form the basis of this thesis.

Of course the other co-authors of my papers—from Düsseldorf, Cambridge, and Mannheim—deserve similar credit. In particular, I would like to thank Florian Jarre from the mathematics department at the University of Düsseldorf. His mathematical skills and ideas were of invaluable help.

I owe gratitude to Jon Crowcroft from the University of Cambridge for providing me with the unique opportunity to visit the Computer Laboratory in Cambridge in spring 2007. It has been a pleasure to work with Wenjun Hu and him during our efforts on blending network coding and congestion control.

Holger Füßler, Matthias Transier, and Marcel Busse from the University of Mannheim not only did a fantastic job at supervising my diploma thesis, but subsequently also became co-authors of my first international research paper. They must thus be attributed the first steps of shaping me into a researcher. Holger in particular has, I guess, never realized how much of a role model he has always been for me. A later collaboration with Matthias finally resulted in the chapter on multicast congestion control in this thesis.

During my time as a doctoral student in Düsseldorf I have (co-)supervised as many as 21 students' theses and 11 master-level student projects—and thus had the great opportunity of frequently working and exchanging ideas with fresh and "unspoilt" minds. Many of the ideas emerging from their work finally found their way into this thesis. In particular this holds for Markus Koegel, Yves Jerschow, and Magnus Roos, whom I want to acknowledge by name here, representatively for many others.

The student helpers in the projects I have been working on also deserve my deep gratitude, for commitment and motivation that exceeded by far what one can reasonably expect for 8 Euros per hour, but also for insightful discussions and unorthodox perspectives. Many of the implementations and evaluations presented in this thesis carry the signature of Alfonso Cervantes and—once again—Markus Koegel.

I also want to thank Marga Potthoff, our group's secretary, for maneuvering me safely through the pitfalls of university bureaucracy, and Christian Knieling, our system administrator, for never complaining about any of my screwy short-term requests.

My parents, Christa and Bernhard Scheuermann, made me to what I am, and they have always unconditionally supported me—thanks for everything! I also thank "Donde" Erika Bradner, for affection and steady support, Peter Steinemann and Sebastian Höfle, for true friendship over many years, and all the other folks back in Mannheim, for always making me feel home.

Finally, I have to admit that I am lacking the words to thank my fiancée, Michaela Metzger, appropriately. For more than ten years now, she has given me something to live for, and supported me in innumerable ways with her love and affection. Thank you.

Contents

Contents

List of Figures

List of Tables

List of Abbreviations

ACK	Acknowledgment
ADTCP	Ad-Hoc TCP
ADTFRC	Ad-hoc TFRC
AIMD	Additive Increase, Multiplicative Decrease
AODV	Ad-hoc On-demand Distance Vector routing
ATCP	TCP for mobile ad hoc networks
ATP	Ad-Hoc Transport Protocol
ATP	Application controlled Transport Protocol
BarRel	Backpressure Reliability
BDP	Bandwidth-Delay Product
BEAD	Best-Effort ACK Delivery
BMCC	Backpressure Multicast Congestion Control
BP	Backpressure Pruning
C^{3}TCP	Cross-layer Congestion Control
CAR	Congestion Aware Routing
CaRe	Capacity Refill
CBR	Constant Bit Rate
CODA	COngestion Detection and Avoidance
COPAS	COntention-based PAth Selection
CPU	Central Processing Unit
CTS	Clear To Send
CWL	Congestion Window Limit
cwnd	congestion window
CXCC	Cooperative Cross-layer Congestion Control
DACK	Delayed ACK

DARPA	Defense Advanced Research Projects Agency
DSR	Dynamic Source Routing
DTC	Distributed TCP Caching
ECN	Explicit Congestion Notification
ELFN	Explicit Link Failure Notification
ENIC	ENhanced Inter-layer Communication and control
EPLN	Early Packet Loss Notification
ESB	Embedded Sensor Board
EXACT	EXplicit rAte-based flow ConTrol
FeW	Fractional window increment
FF	Fast Forward
FIN	Finalize (flag)
FRN	Flexible Radio Network
FTP	File Transfer Protocol
GPS	Global Positioning System
HTTP	HyperText Transfer Protocol
i. i. d.	independent and identically distributed
ICMP	Internet Control Message Protocol
IDS	Intrusion Detection System
IEEE	Institute of Electrical and Electronics Engineers
IP	Internet Protocol
KAL	KeepALive packet
LAD	Least Absolute Deviation
LP	Linear Program
LRED	Link RED
MAC	Medium Access Control
MANET	Mobile Ad-Hoc Network
MLE	Maximum Likelihood Estimator
MMAC	Multicast MAC
NACK	Non-Acknowledgment

NCTS	Negative CTS
noCoCo	Near-Optimal Co-ordinated Coding
NRED	Neighborhood RED
ns	network simulator
NTP	Network Time Protocol
ODMRP	On-Demand Multicast Routing Protocol
OPET	Optimum Packet scheduling for Each Traffic flow
PDA	Personal Digital Assistant
QE	Quick Exchange
RAM	Random Access Memory
RBCC	Rate-Based Congestion Control
RCWE	Restricted Congestion Window Enlargement
RE TFRC	Rate Estimation for TFRC
RED	Random Early Detect
RFA	Request For Acknowledgment
RFC	Request For Comments
RFN	Route Failure Notification
RRN	Route Re-establishment Notification
RSSI	Received Signal Strength Information
RTHC	Round-Trip Hop-Count
RTO	Retransmission TimeOut
RTS	Request To Send
RTT	Round Trip Time
rwnd	receiver advertised window
RWP	Random WayPoint
SACK	Selective Acknowledgment
SCA	Slow Congestion Avoidance
SNN	Sequence Number Notification
SPBM	Scalable Position-Based Multicast
SPBM-BC	Broadcast SPBM

ssthresh	slow start threshold
SYN	Synchronize (flag)
SYNACK	Synchronize + Acknowledge
TACK	Transport layer ACK
TCP	Transmission Control Protocol
TCP-AP	TCP with Adaptive Pacing
TCP-BuS	TCP with BUffering capability and Sequence information
TCP-DOOR	TCP with Detection of Out-of-Order and Response
TCP-F	TCP-Feedback
TCP-RC	TCP-ReComputation
TFRC	TCP-Friendly Rate Control
TPA	Transport Protocol for Ad-hoc networks
TRFA	Transport layer RFA
TSC	Time-Stamp Counter
TTL	Time To Live
UDP	User Datagram Protocol
WSN	Wireless Sensor Network
WXCP	Wireless eXplicit Congestion control Protocol
XCP	eXplicit Control Protocol

Chapter 1

Introduction

This thesis is about information exchange—how to obtain information, how to interpret it, and how to utilize the gained knowledge. It focuses on a specific way of conveying information, on communicating and co-ordinating *implicitly*. It looks at how to learn without asking, how to infer without being told, and how to react without being called on.

We consider wireless multihop networks, or, more specifically, wireless multihop networks with a single channel and omnidirectional antennas. A central, most intriguing property of such networks is that communication is, in some sense, not exclusive between the sender and a single receiver: transmissions are always local broadcasts. Furthermore, the medium locally serializes transmissions: within one neighborhood, only one packet transmission can (successfully) take place at any given point in time. These facts have most often been perceived as a major handicap. They account for effects like spatial instead of per-link bandwidth constraints, or frequent packet losses due to channel variations and radio interference.

Here, we look at the very same properties from a different angle. Often they are, as a matter of fact, not a handicap at all, but instead can form a basis for novel, different solution approaches. It turns out that many challenges in wireless multihop communication can be tackled by seizing the opportunities exactly these networks provide. The key is to use implicit feedback and implicit co-ordination.

The work presented here follows this direction. The subsequent chapters will introduce solutions to a number of central challenges in wireless multihop networking. These solutions have in common that they embrace the media properties, instead of tweaking traditional approaches as an attempt to circumvent the adversarial effects of unsuitable abstractions. This idea as such is not entirely new, it occurs in many previous works in varying shades—the probably simplest and most well-known example are the so-called "passive acknowledgments", where the success of a packet transmission is confirmed when its further forwarding is overheard. But passive acknowledgments are by far not the end of the story, and surprisingly seldom has the concept of availing the medium properties been pursued in all its consequence.

Many of the solutions introduced here can be labeled "cross-layer protocols". Such approaches have often triggered heated discussions: should we really, for the sake of a limited performance improvement, give up a clean architecture? Yet, at least in the context of this work, this probably does not hit the nail on the head. The abstractions made in a network protocol stack, the most widespread specimen today being the Internet protocol stack, are of an intriguing elegance. By generalizing from the peculiarities of specific communication technologies, but likewise of specific applications, layering allows for an otherwise unthinkable flexibility and thus supports the imagination and creativity of technology designers as well as users. This can justly be seen as a major cause for the amazing speed at which networking technology was able to develop over the past decades.

However, while this suggest that the highly successful concept of layering in general should remain unquestioned, it does not mean that the IP stack and its specific layers and interfaces are the one and only optimal solution for all networks. The Internet protocol stack has been designed for a specific network architecture, the Internet. Every protocol designer, regardless how clear the abstractions, makes assumptions on how it works "beyond" the layer currently under consideration—more or less explicitly. The Transmission Control Protocol (TCP), for example, presumes that packet losses are due to congestion, which is clearly a statement on the properties of the underlying network. Wireless multihop networks, however, are in many regards significantly different from the Internet, and TCP's assumption is in fact wrong in this environment. Thus, such assumptions should be subject to careful reconsideration. This alone often suffices to overcome what may otherwise be perceived as major hurdles.

Moreover, since wireless multihop networks will often be closed systems with sufficiently homogeneous devices, strict compatibility of all layers and interfaces to the Internet stack is not necessarily a vital and unquestioned requirement. So, which traditional arrangements should be carried on is to be deliberated, without opportunistically establishing complex interactions and feedback channels "crossing" all layers, but including reorganizations of the responsibilities of functional entities. It is evident that, when doing so, one should always strive for a clarity and elegance on par with the Internet stack. This overall aim in mind, such an approach might also be termed "re-layering".

The main focus of the work presented here is located on what would be assigned to the transport and MAC layers of the Internet protocol stack. Their mechanisms for media access, single-hop reliability, congestion control, and end-to-end reliable transport will be rearranged and reconsidered. The first major contribution, the *Cooperative Cross-layer Congestion Control* (CXCC) protocol, is introduced in Chapter 2. CXCC is based on *implicit hop-by-hop congestion control*, a novel congestion control paradigm which achieves congestion control without explicit co-ordination, in particular without exchanging congestion-related feedback. This is made possible by systematic exploitation of the wireless multihop medium's properties. In CXCC, backpressure towards the source node is established by passively observing the medium. A lightweight error detection and correction mechanism guarantees a fast reaction to changing medium conditions and low overhead.

CXCC provides what could be called "semi-reliable" packet transport, where packets may only be lost in case of failing nodes or links, but not due to queue overflows. In Chapter 3, we will extend the concept of implicit feedback further, to obtain TCP-equivalent, reliable data stream transport. Transport protocols providing such a service typically use end-to-end acknowledgments to guarantee reliable delivery of data. At the same time it is well-known that oncoming control traffic causes serious contention if a shared wireless multihop medium is used. The introduced *Backpressure Reliability* (BarRel) protocol is the first approach ever to be proposed that does not need a continuous stream of acknowledgment packets from the destination. The basic idea is to infer successful end-to-end packet delivery from information obtained locally: due to properties arising from CXCC, the ability to inject another packet into the network can implicitly acknowledge earlier packets. We discuss design choices and introduce two variants of BarRel. The first one requires one single end-to-end acknowledgment at the end of each packet burst. The second variant operates without any oncoming end-to-end control traffic at all.

But the implicit hop-by-hop congestion control concept can not only be extended to reliable unicast communication like in BarRel, it also generalizes to multicast congestion control. We pursue this in Chapter 4, where we introduce *Backpressure Multicast Congestion Control* (BMCC). In mobile ad-hoc networks, the multicast paradigm is of central importance. It can help to save scarce medium bandwidth if packets are to be delivered to multiple destinations. Founded on the idea of implicit hop-by-hop congestion feedback and derived from CXCC, BMCC achieves effective source rate adaption at very low control overhead and packet latency. We put our focus on how to realize it in combination with geographic multicast routing in the Scalable Position-Based Multicast (SPBM) protocol [TFW^{+}07].

Subsequently, in Chapter 5, we will look at an—initially—very different problem in wireless multihop networks. Network coding, after originally having been theoretically discussed primarily for multicast communication in wireline networks, has recently gained more and more attention in practical applications, in particular also in the wireless multihop networking area. Opportunistic network coding has been proposed in [KRH^{+}06] to increase the capacity of the network by combining multiple transmissions in case an intermediate node happens to have matching packets in its queue. However, the spontaneous creation of such situations is left at the mercy of higher and lower layers. We therefore go one step further and study how to create coding opportunities in a more deterministic, yet still practical way. We show how implicit co-ordination can be used to obtain guarantees on the existence of coding opportunities for unicast flows with bidirectional traffic. This leads to a distributed scheduling scheme—*Near-Optimal Co-ordinated Coding* (noCoCo)—that complements CXCC to an astonishing extent.

While most evaluations in this thesis—and in wireless multihop networking in general—are based on simulations, there are also results from a real-world testbed with a CXCC implementation. However, when evaluating such experimental results in detail, like on the packet level, a very fundamental problem arises: the crystal oscillator clocks in the nodes are not very precise, and thus the timestamps in the log files of different nodes are not consistent. As if to confirm the underlying assumptions of this thesis, a solution to the problem once again is possible by exploiting the local broadcast medium. Parallel receptions of the same transmissions can be used to compensate for the deviations of the clocks, yielding globally consistent traces. Our solution thus again makes use of information that can be inferred from the data, without having ever been generated or transmitted explicitly. The proposed method is presented and analyzed in Chapter 6. It is applied after the experiment is completed, using just the set of local log files as its input. It leads to a large linear program with a very specific structure. We exploit the structure to solve the synchronization problem quickly and efficiently, and present an implementation of a specialized solver. Furthermore, we give analytical and numerical evaluation results as well as data from real-world experiments, all underlining the performance and accuracy of the method.

Chapter 2

Implicit Congestion Control: CXCC

The first issue in wireless multihop networking we consider here is congestion control. It has become more and more apparent that wireless multihop networks are much more prone to overload-related problems than traditional wireline networks like the Internet. Appropriate congestion control is thus vital to ensure network stability and acceptable performance.

TCP congestion control is one of the major foundations of today's Internet. It regulates the data rate with an additive increase, multiplicative decrease (AIMD) controller, based on packet loss in the network observed through feedback (or missing feedback) from the receiver. Packet loss is taken as an indication for network congestion. This approach has proven to be highly problematic in wireless multihop networks [GTB99, XS01, dOB02, FML02, FZL+03]. Severe fairness problems, suboptimal throughput and throughput stability issues have been reported. Such effects have also been observed experimentally in real wireless multihop networks, e.g. in [PAM+05].

Raghunathan and Kumar have recently shown that TCP can generally not work as well in those networks as it does in common wired networks, because the rates of multiple TCP flows do not necessarily converge to a fair sharing of the bandwidth due to the shared medium [RK06]. They take this result as "a proof of necessity for a cross layer re-design of TCP+MAC for wireless networks".

All this is not too surprising, considering that the locally shared medium is very different from the full-duplex links that are typical for the Internet. The wireless multihop medium makes congestion a spatial phenomenon: neither nodes nor links, but geographical regions of the network are overloaded. The impact of this can be demonstrated by a very simple simulation experiment. In Figure 2.1(a), ns-2 simulation results of a bidirectional 10-hop chain topology are shown. In the simulation, static routing and the

IEEE 802.11 MAC protocol are used. UDP constant bitrate traffic is injected with increasing data rate at both ends of the chain, traveling towards the opposite end. It can clearly be seen that the obtained total throughput drops rapidly once an optimal load is exceeded. This is due to an increasing number of collisions, leading to more and more packet drops. In a wireline network, throughput degrading effects for a too high UDP load are also well-known; TCP congestion control is able to deal with these problems very well. However, the problem observed here is of a completely different nature, which becomes immediately clear if we set up an equivalent wired topology in ns-2 and measure the throughput in the same scenario, as done in Figure 2.1(b)—due to the duplex connections used between the routers on the Internet, the two packet streams do never even share a link or queue, and thus of course maximum throughput is achieved and maintained.

Note that the observed throughput drop is not a routing effect, since we use static routing without routing overhead and with no link breaks. Enabling 802.11's Request to Send / Clear to Send (RTS/CTS) mechanism does not improve the situation, and is thus not of help here. It is also not a problem of TCP congestion control, since TCP is not used. But TCP is also not an appropriate solution to the problem, as explicated above: TCP-like congestion control doesn't behave well in wireless multihop networks. The results of the experiment demonstrate the spatial nature of congestion in wireless multihop networks. The problem is fundamental, and needs to be taken into consideration by any wireless multihop network design. Thus, the congestion problem in wireless multihop networks deserves to be reconsidered, and there is a need to search for better suited ways to perform congestion control.

In this chapter, we propose a novel congestion control concept for wireless multihop networks, and a concrete protocol design realizing this concept. Both the general approach and the specific protocol also constitute a basis of many of the following results in this thesis. The central results of this chapter have been published in [SLM08].

Our congestion control mechanism is a hop-by-hop approach. Traditionally, hop-by-hop congestion control means that local feedback on the sustainable rate for each node is transmitted to the respective upstream node, in order to establish some kind of backpressure towards the source. Here, we go one step further and actively exploit wireless multihop medium properties to solve problems like congestion control in new ways.

We call the presented approach *implicit hop-by-hop congestion control*, because its foundations are the hop-by-hop nature and implicit feedback, i.e., information gained by

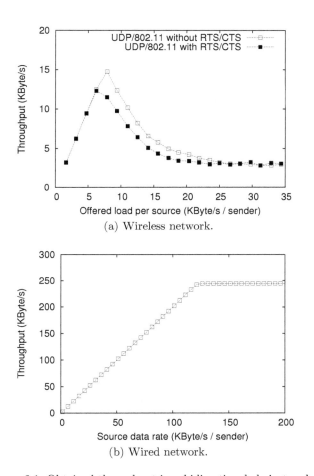

(a) Wireless network.

(b) Wired network.

Figure 2.1: Obtained throughput in a bidirectional chain topology.

observing the transmissions of other nodes in the neighborhood. Even more central, the rate regulation at the source also happens implicitly, just by obeying some simple packet forwarding rules. The protocol design proposed here that realizes the concept of implicit hop-by-hop congestion control is called *Cooperative Cross-layer Congestion Control* (CXCC).

2.1 Related Work

In recent publications several approaches for congestion control in wireless multihop networks have been proposed. Most of them can be classified based on whether they seek to improve TCP or propose alternative approaches. Because of the vast amount of literature especially on TCP modifications for wireless multihop networks, we discuss only those approaches that are more closely related to our work at this point. A much more comprehensive overview of congestion control proposals for mobile ad-hoc networks can be found in Appendix A.

2.1.1 TCP Improvements

Many TCP improvements take the characteristics of wireless multihop networks into account, some examples are [CRVP98, HV99, WZ02, dOB07, FGML02, EKL05]. In the following, we discuss two representatives of this class. For both approaches simulator implementations are publicly available. We use them for comparison purposes in the evaluation of our own protocols.

Fu et al. present Ad-Hoc TCP (ADTCP) in [FGML02]. ADTCP is motivated by a key problem of end-to-end transport protocols in mobile ad-hoc networks: the noisiness of the measurements of indicators for certain network events. To overcome this, different metrics are used: the inter-arrival time of two successive packets, out-of order packet arrivals, the current packet loss ratio, and the short-term throughput in the immediate past. ADTCP combines these metrics in order to obtain a more accurate and robust estimate of the network situation, which then helps to react more appropriately. In ADTCP, the receiver detects the most probable current network state and includes this information into its feedback to the sender.

ElRakabawy et al. propose TCP with Adaptive Pacing (TCP-AP) [EKL05], also an end-to-end approach. In TCP-AP, the focus is on avoiding large packet bursts. For this purpose, the packets that are allowed to be sent out by the TCP congestion window are paced adaptively. The authors define the 4-hop propagation delay as the time between the transmission of a packet by the TCP source node and its reception by the node four hops downstream. TCP-AP estimates this value from the round-trip time (RTT) of the packets. In combination with the coefficient of variation of the RTT samples the estimate is used to establish a minimum time between two successive packet transmissions.

2.1.2 Alternative Approaches

Our own approach to congestion control, CXCC, clearly falls into the category of alternative approaches. Therefore, we investigate the work in this area more closely.

The Ad-hoc Transport Protocol (ATP) by Sundaresan et al. [SAHS03] is a rate-based, network supported transport protocol for mobile ad-hoc networks with end-to-end congestion control. The authors consider TCP's mechanisms inappropriate for ad-hoc networks. Thus, ATP is designed as an "antithesis" to TCP. ATP strictly separates congestion control from reliability mechanisms and requires only limited feedback from the receiver. The intermediate nodes piggyback the maximum queuing delay along the route on the packets passing by. This information is then used to determine the appropriate rate at the source node. A similar approach to ATP is Explicit Rate-Based Flow Control (EXACT) by Chen et al. [CNV04]. In EXACT, not delays are transmitted, but the intermediate nodes calculate sustainable rates for each flow directly; these are then piggybacked. To accomplish this, EXACT, in contrast to ATP, requires state information for each flow in the intermediate nodes. Both approaches show that effective congestion control can be performed in different ways than with TCP, and that it can be tailored for mobile ad-hoc networks. However, both approaches control the rate only at the source node, based on feedback from within the network. The time until the feedback eventually arrives at the source node is relatively long, considering the rapidly varying medium conditions. By using in-network reaction to congestion, our scheme guarantees an immediate, localized adaptation to a changing environment.

Yi and Shakkottai [YS07] discuss the usage of hop-by-hop congestion control for wireless multihop networks. Their theoretic approach shows the feasibility of hop-by-hop congestion control in these networks. It provides a good basis for the theoretical understanding of the behavior of congestion control feedback schemes. From a more practical point of view, however, it has to be considered that the assumptions they make are not fulfilled by today's common wireless hardware. The proposition that in a wireless network any two links not sharing a common node are completely independent from each other cannot be met. Additionally, their approach uses explicit feedback to the upstream nodes, and therefore imposes load for the feedback traffic on the network. In contrast our scheme avoids additional load during congestion situations.

A concept that is in some aspects similar to ours has been used in the DARPA packet radio network with the "adaptive pacing" protocol [GJ82]. Packet radios using this protocol wait for a "passive acknowledgment" before the next packet is transmitted along a link. This is combined with rate throttling based on delay measurements. For these measurements the protocol requires to maintain state information for each neighbor of a packet radio, which will quickly become outdated. Thus, we assume that the central aspect of their work—the measurement and calculation of inter-packet delays (or rates, accordingly)—cannot be immediately adopted for wireless multihop networks with IEEE 802.11-like physical layers. Our simulations underline this assumption.

"Passive" or "implicit" acknowledgments obtained by overhearing have been mentioned in several publications since they have been introduced for packet radio, for instance in the context of the Dynamic Source Routing (DSR) protocol [JM96]. In our approach, implicit acknowledgments are one central element of the congestion control mechanism.

Zhai at al. propose an approach which they call Optimum Packet scheduling for Each Traffic flow (OPET) [ZF06]. It consists of four mechanisms to reduce the impact that wireless medium contention has on throughput and fairness in Mobile Ad-Hoc Networks (MANETs). One is a hop-by-hop backpressure scheme, similar to both packet radio adaptive pacing and our approach. Their mechanism is used in combination with standard TCP, and the backpressure mechanism is tightly coupled to the RTS/CTS mechanism in 802.11, using explicit "Negative CTS" packets. While both design decisions are reasonable when compatibility must be preserved, we chose a more radical approach. We replace TCP altogether and thus avoid any problems of possible feedback loops between multiple congestion control schemes on different layers of the protocol stack. Furthermore, our scheme is able to send fewer explicit messages and thus minimizes the load on the network.

Hop-by-hop congestion control has been a basis of several proposals in the context of wireless sensor networks (WSNs). In WSNs, typical traffic consists of relatively small packets and is directed to or from a small number of special nodes, the sinks. Thus, the sensor network approaches have in common that they consider packet flows that are directed to or from such a sink, whereas we primarily consider unicast traffic among arbitrary pairs of nodes. A survey on congestion control in sensor networks is provided in [WSL+06].

In backpressure-based congestion control approaches for WSNs, congestion feedback is typically either communicated via dedicated packets like in Wan et al.'s Congestion Detection and Avoidance (CODA) protocol [WEC03], or piggybacked as for example in Hull et al.'s scheme named Fusion [HJB04]. A general problem with both approaches is that it requires packets to be sent in order to inform the neighbors about the congestion situation. If a node cannot send a packet due to an occupied medium, it will not be able to signal congestion. This problem does not exist in our scheme.

Implicit acknowledgments have also been used in the WSN context, e.g., by Woo and Culler [WC01], with the primary purpose of reducing the control packet overhead. Also in [WC01], an AIMD-based backpressure rate control approach is proposed. The nodes maintain a probability with which a packet is allowed to be sent out or forwarded, essentially determining the aggressiveness of the medium access. The probability is based on the observation of the downstream node's forwarding behavior. This yields congestion feedback that propagates back towards the sources. In our scheme, no windows, rates or forwarding probabilities need to be maintained.

Since no dedicated congestion messages are used, schemes like [WC01] and [HJB04] are also often called "implicit". Note, however, that our notion of implicit here is slightly different: in our work, "implicit congestion control" means that the congestion control works by taking advantage of the medium properties in various ways. In particular, it is performed without explicitly maintaining window sizes or packet rates, without any node explicitly deciding whether there currently is congestion or not, and without any explicit congestion feedback, be it in dedicated packets or piggybacked.

2.2 Algorithmic Idea

2.2.1 Shared Medium Model

To motivate the implicit hop-by-hop congestion control approach we introduce a very simple model for the effects of the shared medium. It is easy to see that in any part of a network—wired or wireless—, on a sufficiently long time scale to avoid short-term effects, the output rate of traffic forwarded through this area (OUT) cannot exceed the forwarded traffic input rate (IN). We denote this fact by

$$\text{OUT} \leq \text{IN}. \tag{2.1}$$

In common wireline networks, there are also separate, independent upper bounds for the input and output data rates, given by the bandwidth of the respective links. If we denote the total ingress link bandwidth into the area under consideration by BWIN, and similarly the total outgoing bandwidth by BWOUT, we have

$$\text{IN} \leq \text{BWIN} \tag{2.2}$$
$$\text{OUT} \leq \text{BWOUT}. \tag{2.3}$$

The important point is that, in this constellation, it is not of immediate disadvantage for the throughput if the input rate exceeds the output rate. Of course this will lead to dropped packets, but these drops will occur *before* the bottleneck—in this case the output—, and maximum throughput will still be obtained.

This is the foundation of any common end-to-end congestion control. TCP, but also, e. g., the TCP-Friendly Rate Control (TFRC) protocol [FHPW00], utilize packet drops to determine the bottleneck bandwidth. To accomplish this, drops are actually provoked: TCP congestion control aims to keep the buffers in the network full, causing strict inequality in (2.1).

If we consider the situation in a wireless multihop network, the shared medium adds a new aspect to the above considerations. Instead of independent bounds on the input and output rate as in (2.2) and (2.3), ingoing and outgoing links within a collision domain share the available bandwidth. This leads to a constraint of the form

$$\text{IN} + \text{OUT} \leq \text{BW}, \tag{2.4}$$

where BW is the abovementioned shared total bandwidth.

This has an important implication: here, the output rate cannot be optimal in the case of strict inequality in (2.1). Therefore, the situation for the congestion control is completely different. Due to (2.4), increasing the input rate beyond the output rate will result in a suboptimal overall throughput. This is one key reason for TCP's fundamental problems in wireless multihop networks.

These observations can also serve as an explanation for the previously described throughput breakdown in case of network load beyond the optimal point. Excessive network input will yield a more and more extreme inequality in (2.1), an increasingly dominant role of the input in (2.4), and consequently leads to decreasing throughput.

Implicit hop-by-hop congestion control as proposed here is based on a completely different feedback paradigm than TCP. Our approach aims to achieve strict equality in (2.1) even on a short-term time scale. The key observation is that this is actually possible in wireless multihop networks, by actively exploiting the local broadcast property.

2.2.2 Implicit Hop-by-Hop Congestion Control

We define a flow as a directed pair of communicating nodes, i.e., all the packets traveling from a node A to another node B belong to the same flow. Other definitions are possible, but since neither the idea of implicit hop-by-hop congestion control nor the CXCC protocol necessarily depend on this definition this is, for now, only a matter of form. With implicit hop-by-hop congestion control, the protocol enforces that the input rate for a given flow does not exceed the output rate at any intermediate node. This is accomplished by preventing the transfer of a second packet to a node until this node has forwarded the previous one. Every node along the route thus queues at most one packet of a flow, and no further packet is forwarded until the queue space at the next node is free again. We call a flow *blocked* in a node when there is no space for the next packet available at the downstream node.

The mechanism of waiting for the next hop to forward one packet before passing on the next one leads to a fast and efficient implicit backpressure mechanism towards the packet source. If a node is not able to forward a packet immediately, it will thereby implicitly stop the input flow from its predecessor, and so on, until the packet source itself will not be allowed to send the next packet. The concept ensures that the input rate cannot exceed the output rate, and it yields a very fast reaction to changing medium condition. If the forwarding is delayed, the backpressure is established immediately. Furthermore, such a mechanism works for both TCP-like and UDP-like traffic, whereas TCP congestion control as well as most other approaches necessarily depend on reliable transmission and corresponding end-to-end feedback. In our approach, the backpressure is independent from end-to-end reliability mechanisms.

Another possible interpretation of implicit hop-by-hop congestion control is that a node is only allowed to transmit if there is a "hole" at the downstream node. Thus, these holes propagate along the route in reverse direction. Before a network bottleneck, where backpressure builds up, holes are rare. The nodes there are not allowed to transmit until a hole has propagated through the bottleneck area; therefore, the input into the congested area is limited. Behind the bottleneck, the holes dominate, meaning that the output of the congested area is not constrained.

For the backpressure mechanism to work, a node has to know whether the downstream node still has a packet of the same flow in its queue. This information could be sent explicitly, but such a mechanism would induce additional traffic and thus potentially aggravate congestion situations. In our approach, the shared nature of the medium is exploited to gain the necessary information at no additional cost. In most wireless technologies that are considered as a basis for multihop communication, transmissions are de-facto broadcasts: the radio waves propagate to all nodes in the vicinity of the sender, in particular also to the upstream node. Then the forwarding of a packet by the downstream node can be overheard. Here, we consider this kind of network. The central rule of implicit hop-by-hop congestion control is then very simple: if the forwarding of the previously sent packet is overheard, this indicates that the next one may be transmitted (Figure 2.2(a)).

At the same time this implicit notification about a free queue space can—as a side effect—serve as an implicit acknowledgment, indicating successful packet delivery to the next hop. Note that this requires no additional, piggybacked header fields. If it has a unique ID, the packet is sufficient as-is. Common MAC layer acknowledgments are therefore no longer required. This is also called "passive acknowledgment" in the literature.

Due to backpressure or medium contention, the forwarding can be delayed. Thus, the nodes have to tolerate a significant delay before the implicit acknowledgment. We don't see this as a problem; instead, we accept this kind of delay tolerance as a necessity for robust wireless multihop protocols. This leads to another key concept of our approach, which we call *soft timing*. Due to the local broadcast property of the wireless medium it is very hard to predict when a chance to transmit a packet will arise. Thus, it is also not unrestrictedly possible to guarantee a collision-free answer from a specific node within a certain, tight time frame. After all, the medium around this node might be busy, preventing it from sending. 802.11's way to deal with this issue is to ignore carrier sensing when transmitting MAC layer acknowledgments; however, this can cause additional collisions and therefore wastes bandwidth. We thus advocate to use asynchronous answers within a comparatively long period of time—potentially only after several other transmissions have happened.

Obviously, when a packet reaches its final destination node, there will be no further forwarding. Thus, the packet cannot be implicitly acknowledged. The destination node therefore needs to send an explicit acknowledgment. By delaying this acknowledgment if the packet cannot be immediately passed up the protocol stack, an integration with flow control is possible—that this is tolerable for the protocol can be seen as a consequent generalization of the soft timing principle.

It is conceivable to allow a higher number of unacknowledged packets to be transmitted before a node has to wait for an acknowledgment and backpressure builds up. However, the medium enforces a serialization of all local transmissions. Note that, as a consequence of this medium property, one single unacknowledged packet suffices to fully utilize the available bandwidth. This fact is also confirmed by the simulation results presented below. Thus, while allowing for more than one unacknowledged packet largely increases the complexity of the protocol, it does not yield substantial benefits. In fact, additionally allowed queue spaces for unacknowledged packets would be occupied by packets primarily before a bottleneck, but not behind the congested area, where packets can be forwarded away quickly. This increases the backlog in front of the bottleneck, the queuing delay and, consequentially, the packet latency, and is therefore highly undesirable. But it does not allow the packets to be forwarded any more quickly *through* the bottleneck.

2.2.3 Deadlock Freeness

In implicit hop-by-hop congestion control, flows are stopped completely when backpressure occurs. Thus, it is crucial to verify that this blocking will eventually be released, i. e., that no deadlocks can occur where flows are blocked indefinitely—e. g., because they wait for each other. However, it is easy to see that our scheme will not run into such a deadlock situation.

When a queue is blocked in a node, then this is because it waits for the next node downstream to forward another packet of the same flow. This node might in turn wait for the next node downstream, and so on. Eventually, the queues will, directly or indirectly, all wait for the destination node of the flow. Thus, as long as the destination node is accepting packets, the blocking will be released.

In order to formalize and generalize this statement, we define the *wait graph* of a network. The wait graph is the directed graph where the vertices are the queue instances in all nodes, and where a pair of queues (q_1, q_2) is in the set of edges if and only if q_1 is currently waiting for q_2. If the wait graph is loop-free, then there is no deadlock situation: all queues will eventually wait either for a non-blocked queue or for the destination node. In our scheme, the wait graph is always loop-free if the routing is loop-free. This holds because a queue q_1 of flow f in node n_1 can wait—directly or indirectly—only for another queue q_2 if q_2 also belongs to f and is located at a node n_2 such that n_2 is further downstream on f's route.

This is the reason why implicit hop-by-hop congestion control works on a per-flow basis. One could also think of a scheme where a node generally waits for the next node to forward the previously sent packet before the next one is transmitted, regardless of the flow it belongs to. In this case, however, the loop-freeness of the wait graph would not be ensured.

2.3 Basic CXCC

Now that the general concepts of implicit hop-by-hop congestion control have been reasoned and explained, we focus on how to realize this idealized, abstract idea in a real network, where packet loss due to collisions and other adversities is common. In order to do so, we will describe a first, basic version of our CXCC protocol.

2.3.1 Dealing with Lost Packets

In wireless multihop networks it is common that a packet transmission is unsuccessful. For implicit hop-by-hop congestion control packet loss poses a serious threat: a packet loss will block the transmission of data for the flow that the lost packet belonged to. Thus, CXCC needs to be able to recover from packet loss efficiently.

The most basic loss situation is that the next hop node does not receive a data packet transmission (Figure 2.2(b)). No more packets would then be forwarded: since the next hop will of course not forward a packet it has never received, no implicit acknowledgment will arrive. The simple solution to this problem in the basic CXCC protocol is to repeat the packet transmission if, after a certain timeout, no acknowledgment has been received.

A data packet transmission is not only used to communicate the data in downstream direction, but also to acknowledge the reception of the packet at the same time implicitly. The situation that the upstream node, waiting for the acknowledgment, does not receive the packet is also possible (Figure 2.2(c)). The upstream node will then stop sending any further packets because it has missed the implicit ACK. Note that the upstream node is not able to distinguish this situation from the case above: it cannot tell if the

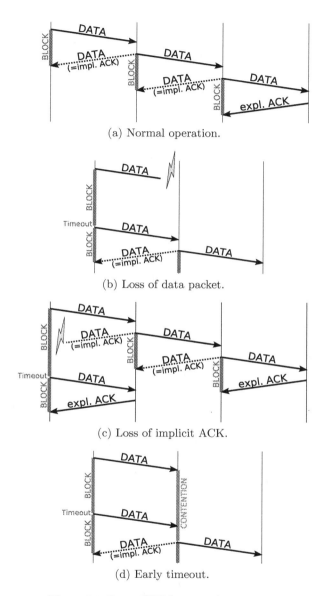

(a) Normal operation.

(b) Loss of data packet.

(c) Loss of implicit ACK.

(d) Early timeout.

Figure 2.2: Basic CXCC protocol operation.

next hop has not received the packet, or if the implicit acknowledgment has not been received. It will thus conduct a packet retransmission as described above.

The packet had, however, already been successfully forwarded, so this could lead to packet duplication. Therefore, the proposed solution has to be augmented by a duplicate detection mechanism. This is easy to accomplish, since only a duplicate of the *most recently* received packet is possible. When a second copy of a packet is received, the next hop will drop the duplicate and will not forward the packet again. However, then the previous hop will again not receive an implicit acknowledgment. To overcome this, we propose the following behavior: a node sends an explicit acknowledgment when it receives a duplicate of a packet it has already forwarded and for which it has already received an (implicit or explicit) acknowledgment. If these conditions do not apply it silently ignores the duplicate. This ensures that an ACK is only sent when there will definitely be no further chance to acknowledge the packet implicitly. This avoids unnecessary transmissions.

There is a third situation that can also not be distinguished from the ones above by the upstream node: the timer of the upstream node could expire before the next hop has been able to forward the packet. This is depicted in Figure 2.2(d). In this case, the next hop node will have the packet in its queue and it will ignore the received duplicate. It will not transmit any explicit information to the upstream node. The reason for this behavior is twofold: first, an explicit acknowledgment would not be of any direct use for the upstream node, since its reaction would be the same as if nothing at all is received: it would wait. Second, if the packet is still in the queue this probably means that there is network congestion. Thus, if there is a chance to transmit *anything* at all, the precious medium time should be used to transmit data—which will also serve as an implicit acknowledgment—rather than an explicit control packet.

2.3.2 Queuing in CXCC Nodes

The forwarding rules established above define how a basic CXCC node works. An important implication of these rules is that queuing in each node requires some attention. For each flow passing through a node a queue has to be maintained. These queues can be created on-demand when the first packet of a stream arrives. Because of the one-packet-per-hop constraint the queue has to provide space for at most two packets: one that has already been forwarded but is not yet acknowledged, and therefore needs to be cached in order to be retransmitted if necessary. The other one that has been received from the upstream node but must not be forwarded until an acknowledgment for the preceding packet arrives from the next hop node.

As one node can handle queues for different streams, it is necessary to decide out of which queue a packet shall be forwarded. Normally, one of the queues that are not empty and not blocked is chosen randomly. Retransmissions after a timeout are handled just like the first transmission of the respective packet. If an explicit acknowledgment is waiting to be sent, it is given priority over data packets. More sophisticated schemes could be integrated to enforce, e. g., certain quality of service or fairness metrics.

From the discussion above it can be seen that CXCC requires to keep per-flow state in the intermediate nodes, and that it also implies a certain computational overhead. There are per-flow queues, and the duplicate detection mechanism needs to remember the last packet of each queue for some time. However, we do not expect this to be a problem in practice. For a node in, e. g., a mobile ad-hoc network, one can expect a relatively high computational power and many resources in relation to the small effective bandwidth of the shared multihop medium. Because of this small bandwidth the number of flows crossing one node is also quite limited, compared to, for example, typical Internet routers.

Furthermore, the information does not need to be kept for a long time. When no more packets of a stream are queued in a node, the corresponding queue can be removed immediately. The information that is needed for the detection of duplicates can also be removed quickly, since it can reasonably be expected that a duplicate can only occur within a relatively short time. So, the small additional overhead is perfectly reasonable and does not significantly limit the scalability of our approach.

2.3.3 Retransmission Timeouts

The choice of an appropriate retransmission timeout is a crucial factor for CXCC to work properly. In CXCC, the timeout TO before a packet retransmission is scheduled depends on the packet's size and thus the medium time needed for the transmission of a packet. The packet transmission time T_P can be easily calculated in the case of a constant medium bandwidth B, since the packet size s_P is known:

$$T_P = \frac{s_P}{B}. \tag{2.5}$$

Here, we use an exponentially increasing timeout between consecutive unanswered transmissions. It is calculated in the following way:

$$TO = \alpha^r \cdot D \cdot j \cdot T_P, \qquad (2.6)$$

where the exponential base α will typically be between 1 and 2 (we use $\alpha = 1.2$ here). r increases by one with each retransmission without feedback, and is reset to zero after the reception of an ACK. D is a constant (here, we use $D = 3$), and j is a random factor that adds some jitter to the transmissions (we choose j randomly in the interval $[0.975, 1.025]$ here). Simulation experiments indicate that CXCC is quite robust regarding the exact values of the parameters; varying them has only limited impact on the performance.

Generally, the timeouts resulting from the above formula might seem like a very short time. But an elapsing packet retransmission timeout does not imply an immediate retransmission. It just *allows* the retransmission of the packet. It still has to wait until the medium is free and a packet can actually be sent, and until the respective queue is chosen out of potentially several alternatives. Thus, the actual delay until a retransmission is performed depends largely on the network conditions and current media utilization; the retransmission timeout just establishes a lower bound on this time. Therefore, the scheme exhibits some inherent adaptivity. Note in particular that it also remains possible for a packet to be acknowledged after the timeout has elapsed, while it is waiting for its retransmission.

2.4 Request for ACK

2.4.1 First Simulation Results with Basic CXCC

As shown in Figure 2.1 the throughput of an 802.11-based wireless multihop network goes up with increasing data rates, but beyond some optimal point it cannot maintain this throughput. Instead, due to collisions and retransmissions, the throughput decreases rapidly. Basic CXCC in the same topology on the other hand exhibits one important characteristic: it is able to stabilize the throughput at a reasonable level if too high input data rates are offered. Figure 2.3 shows this trait of CXCC.

These results are on the one hand very promising, as the basic CXCC protocol is able to handle the congestion situation and to guarantee stable throughput over multiple

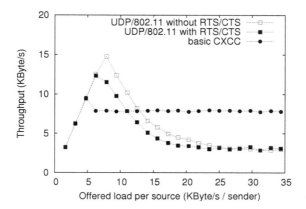

Figure 2.3: Basic CXCC throughput in a bidirectional chain topology.

hops. However, the achieved throughput level is significantly lower than the maximum achieved by UDP at the optimal input data rate. In the following, we will introduce an improved version of the CXCC protocol that addresses this problem of the simple basic CXCC scheme.

The reason for the suboptimal performance lies in the packet retransmissions as they are performed in basic CXCC. Whenever a node does not receive an acknowledgment and a timeout expires, the whole data packet is retransmitted. However, if only the implicit acknowledgment has been lost, or if the next hop node has simply not yet been able to forward the packet due to contention or backpressure, it is unnecessary to retransmit the whole packet: it has already been successfully transmitted before, the payload has already arrived at the next node.

2.4.2 RFA Mechanism

In Section 2.3.1, three situations have been distinguished where basic CXCC performs a packet retransmission. It has also been reasoned that the retransmitting node is not able to tell these situations apart. But only in one situation it is actually necessary to retransmit the payload. Therefore, a strategy is missing that helps to avoid unnecessary transmissions. For the development of such a strategy it is important to know which of the three loss situations is the dominating one. We conducted simulations to obtain this information. Figure 2.4 shows which of the three loss situations causes how many retransmissions in each of the nodes of a simulated bidirectional 10-hop chain.

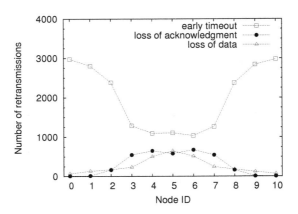

Figure 2.4: Reasons for packet retransmissions in bidirectional chain topology.

From this figure it becomes clear that, at least in this topology, the cases where a payload retransmission is *unnecessarily* performed—when an acknowledgment is lost or when an early timeout occurs—clearly outnumber the situations where a data packet has been lost and a retransmission is thus needed. Results from other topologies and the performance figures presented later on point in the same direction. One possible interpretation of the distribution of the three retransmission reasons is that the timeouts introduced in the previous section are too short, resulting in too many or too fast retransmissions. However, our simulations also show that longer retransmission timeouts do not improve the performance of basic CXCC, because then the recovery from true loss situations happens with a too long delay. Therefrom arises the necessity for a scheme that is able to recover fast enough from a real loss situation, but at the same time avoids the large number of unnecessary packet retransmissions in the case of delay through backpressure.

Our solution to this problem is to not retransmit the whole data packet after an expired timeout, but instead just a small control packet. We call these control packets *Request For Acknowledgment* (RFA). An RFA contains all the important header information from the data packet, but not the payload. It thus provides the downstream node with all the information that is necessary to react appropriately.

We now look at the reaction of the downstream node upon reception of an RFA packet. The simplest case is when the transmission of the data packet had not reached the next hop, i.e., the RFA refers to an unknown packet. Then, a retransmission including the payload is necessary, and should happen as soon as possible. If the downstream node detects this situation, it thus tells the upstream node to retransmit the full packet by sending an explicit negative acknowledgment (NACK) frame. This is depicted in Figure 2.5(b). Because of the additional RFA-NACK-handshake the overhead in this case is actually higher than with basic CXCC. However, as the previously stated simulations indicate, this occurs rather seldom. In the other two cases, a retransmission of the data packet is not necessary, and thus a lot of otherwise unnecessarily occupied medium time can be saved with the RFA scheme.

When just the implicit acknowledgment has been lost although the packet had actually been correctly forwarded, the downstream node sends an explicit acknowledgment if the packet has been forwarded further and acknowledged by *its* downstream node, as explicated above. This behavior, shown in Figure 2.5(c), is the same as for basic CXCC.

In case of an early timeout, the reaction is also identical to that of basic CXCC: no action is performed, as shown in Figure 2.5(d). Therefore, the only significant difference in this as well as in the previous case is that the payload has not been retransmitted and thereby medium time has been saved.

It should be noted that for an upstream node an overheard RFA packet can serve as an implicit acknowledgment for the packet it refers to.

2.5 Dynamic Routing: Detecting Broken Links

So far we have considered situations with stable routes. In particular, we have assumed that each next hop node—even though single transmissions to it may fail—is generally reachable. In a mobile environment, where nodes move and the topology changes continuously, it is, however, necessary to be able to detect when this is no longer the case.

Different ways have been put forward how a wireless multihop routing protocol can detect that a neighbor is no longer reachable. One way to accomplish this is via periodic beacon packets. A node is reachable as long as the exchange of beacons is successful. This is common in proactive routing approaches, and had, e.g., been followed in the original Ad-hoc On-demand Distance Vector (AODV) routing paper [PR99]. Beacons, however, constitute additional network traffic, and due to the limited beaconing frequency such an approach reacts rather slowly. A different way to tell when links are

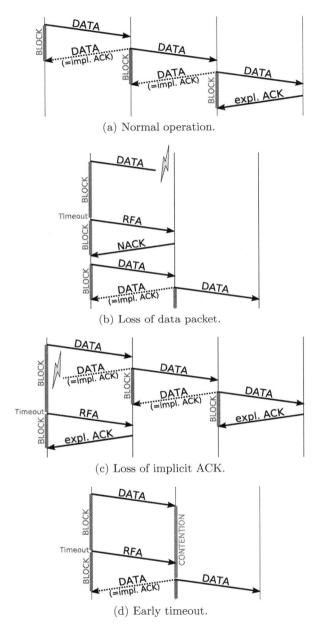

(a) Normal operation.

(b) Loss of data packet.

(c) Loss of implicit ACK.

(d) Early timeout.

Figure 2.5: Protocol operation with RFA control packets.

no longer working is thus commonly used especially in IEEE 802.11-based settings. In 802.11, there is a limit on the number of retransmissions of a frame. If this limit is exceeded and still no link layer acknowledgment is received, a link error is reported to the routing protocol.

In a CXCC-based network, the point of departure is slightly different, since there are no 802.11-like link layer acknowledgments. Transferring the idea is thus not straightforward. If RFAs are used as introduced, reporting a link break after a fixed number of unanswered RFAs is surely not wise. Then a route might wrongly be considered broken in case of heavy backpressure: as seen before, the next hop node ignores RFAs if the packet has arrived, but could not yet be forwarded.

To overcome this, we extend the CXCC protocol again. Note that this extension is only necessary in case the discussed kind of link status feedback is desired or needed.

In the variant of CXCC with link status feedback, a node may send a *keepalive* (KAL) packet, when it receives an RFA for a packet that is held back due to backpressure. Thereby, it signals that the link is working, without releasing the backpressure. After receiving such a KAL packet, a node will continue to send RFAs—potentially again answered by KALs—in order to verify that the link stays up. To save bandwidth, this can happen at a lower frequency. After all, if a KAL is received, there is backpressure; thus, a congestion situation is likely, and bandwidth is a scarce resource. When we use KALs in this work, we accomplish the slowdown of the RFAs by increasing the constant D in the timeout (2.6) from 3 to 5 after the reception of a KAL. By using KALs, a link may safely be considered broken if too many consecutive RFAs without any answer occur. We will use a threshold of seven consecutive unanswered RFAs here, resembling the seven unsuccessful RTS retransmissions typically used in 802.11.

Many researchers have already observed that always assuming a link failure in case of an 802.11 transmission error callback can lead to a large number of spurious route breaks; see, e. g., [EKL05, DB04, PPW⁺07]. Our results suggest that a main reason is the aggressive retry behavior of 802.11, which performs RTS or data retransmissions after a very short delay and expects an immediate reply. The soft timing principle in CXCC and the exponentially increasing timeouts between consecutive RFAs turn out to have very beneficial effects here. CXCC's RFAs are sent at longer intervals than standard 802.11 RTS frames, and answers may arrive with significant delay. Consequently, there is much more opportunity for the destination node to reply, especially if, e. g., the medium around it is currently busy. This largely reduces the number of spurious route errors, and thereby also the amount of unnecessary routing traffic in the network.

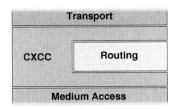

Figure 2.6: CXCC's position in the protocol stack.

2.6 Layer Interfaces

While the previous sections dealt with the CXCC protocol itself and its functions, we now discuss how it can be integrated into a protocol stack. It uses modified interfaces in particular to the MAC and routing layers. We are aware that such a design requires significant control over the network participants' protocol stack. However, our point is that the cost of preserving an unmodified Internet protocol stack is too high. Since wireless multihop networks can often be expected to be homogeneous, and potentially even tailored for one specific application, we consider the required modifications feasible.

An overview of the position of CXCC in the protocol stack is shown in Figure 2.6. The congestion control function is moved from its traditional position in the transport layer down to the CXCC module. Therefore, the transport layer protocol can focus on its main tasks: providing process-to-process communication (i.e., ports), and, optionally, byte-stream-based, connection-oriented communication primitives, preservation of packet ordering and/or reliability. Note that this does not necessarily mean a different interface to the application: a TCP/UDP-compatible socket interface to the application layer can be provided if CXCC is used, meaning that existing applications do not need to be changed. A transport protocol design that fits very well with CXCC will be presented in the next chapter.

CXCC does not interfere with the core functionality of the routing layer. This is important in order to stay independent from a specific routing approach. CXCC essentially encapsulates the routing protocol. This is necessary since CXCC components need to reside both above and below the routing protocol. In a practical implementation CXCC can take over the responsibility for packet queuing completely. This simplifies the maintenance of the per-flow queues, and the routing protocol can concentrate on its most basic functionality: providing information on the next hop towards a given destination node. This implementation aspect is, however, not mandatory, and it is independent from CXCC's congestion control functionality.

Finally, the MAC layer is also reduced to its core responsibility: observing the medium and deciding when a transmission is allowed to take place. There is no need for link layer retransmissions if CXCC is used. Such mechanisms have been introduced in wireless MAC protocols to overcome the inherent unreliability of the medium. However, CXCC is aware of the medium properties and does not need this support. Retransmissions and acknowledgments are handled by CXCC itself in a very efficient manner. Likewise, CXCC does not use the RTS/CTS mechanism for virtual carrier sensing. Our simulation results show clearly that RTS/CTS has in fact a mostly negative performance impact in wireless multihop networks. This observation is in accordance with many previous results, e. g., in [XS01, XGB02]. RTS/CTS has been designed for single-hop wireless networks, and does not fully solve the hidden terminal problem in a multihop environment, but instead causes new problems, like false blocking due to overheard, but failed RTS/CTS handshakes [RS07].

One additional difference between the protocol stack of a CXCC node and an Internet node is a changed interface to the MAC layer: there exists no interface queue between CXCC and the MAC. Instead, the queuing is handled by CXCC. The MAC provides feedback on when a packet may be sent. The reason why we remove the common interface to the MAC layer is easy to see: with an interface queue, it would be possible that a packet is enqueued, but can not be sent immediately. It thus might happen that before the transmission can be started, other messages are received that redundantize the waiting transmission—like an acknowledgment for this packet. In this case an unnecessary transmission would be performed, even though the transmitting node actually has the knowledge that it is of no benefit. Our design avoids such effects. Another approach would have been to provide means of altering the interface queue upon demand. However, this would require a much more complex interface, and thus contradicts our intention of keeping the functional separation clean.

Due to the MAC modifications CXCC can not immediately be used with most commodity wireless hardware available today. This is because typically significant parts of the MAC implementation are located in the firmware and cannot easily be changed. However, the firmware modifications that are necessary in order to allow using CXCC on existing hardware are limited. And, as our results show, the possible gain is significant. More importantly, many, if not most, wireless multihop networks will be application-specific, and will often be based on tailored hardware.

2.7 Simulations

In order to examine the performance of CXCC we have performed an extensive simulation study, using the network simulator ns-2 [ns2a]. The evaluation is based on different scenarios. First, we compare the behavior of CXCC and that of UDP traffic over IEEE 802.11 in three different topologies, where packets are produced by constant bit rate (CBR) traffic sources with increasing frequency. This provides us with some general insights on how well CXCC is able to adjust a source's rate in order to utilize the capacity of the network. Thereafter, we compare CXCC to three other congestion control protocols: TCP Newreno [FHG04], ADTCP [FGML02], and TCP-AP [EKL05]. TCP Newreno is the most commonly used TCP variant today; we use the implementation from standard ns-2. ADTCP and TCP-AP are end-to-end approaches that aim to improve TCP performance in mobile ad-hoc networks. For our comparisons, we use the respective protocol implementations that have been made available by the authors.

In all simulations presented here the packet payload size is set to 512 bytes. We have also performed simulation runs with smaller and larger packets. Although the absolute results were of course different, the overall outcome remained the same. The network bandwidth for all simulations is fixed to one megabit per second. The two-ray ground radio propagation model is used with the common settings of 250 meters radio range and 550 meters carrier sense radius. Further results on CXCC's performance, including, e. g., simulations with a different radio propagation model and evaluations of CXCC's resilience to bit errors on the wireless medium, can be found in [SLM08].[1]

We use static routing with optimal routes (with regard to the hop count), in order to remove any influence of a routing protocol and to focus solely on the inherent challenges of congestion control in a wireless multihop network. In a static setting, route breaks do not occur, and CXCC's retransmissions and RFAs can guarantee single-hop reliability. Therefore, end-to-end reliability is ensured in our CXCC simulations, without additional measures. Results with implicit hop-by-hop congestion control in conjunction with dynamic routing protocols will be presented in subsequent chapters.

---

[1]The results presented in [SLM08] and here exhibit the same overall picture, but differ slightly in details. This is due to a different parametrization of the protocols and the simulated topologies. In particular, the node spacing in the deterministic topology simulations and details of the timeout mechanism in CXCC differ.

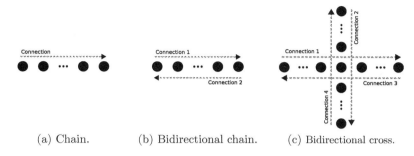

     (a) Chain.   (b) Bidirectional chain.   (c) Bidirectional cross.

Figure 2.7: Deterministic simulation topologies.

2.7.1 Deterministic Topologies

In our simulations with deterministic topologies and fixed packet streams we consider the three different scenarios shown in Figure 2.7. All of them are based on 10-hop equidistant chains, with neighbored nodes placed 150 meters apart. We have also performed simulations with different chain lengths and obtained very similar results.

The first and second topology are based on one of these chains. Along this chain, we use one single connection from the first to the last node in the chain as our first scenario ("chain"). In the second scenario, we added a second stream of packets to the same topology, running from the last towards the first node, that is, in the opposite direction ("bidirectional chain"). The third scenario is based on two chains, one in x, the other one in y direction of a plane. They cross each other in the middle of the chains, where they share one common node. "Bidirectional cross" has four data streams altogether, one starting at each chain end, with their destinations at the opposite end of the respective chain.

Like in the bidirectional chain simulations described before in the introduction to this chapter and in Section 2.4.1, the CBR sources at the end of the chains were configured to produce packets at a fixed data rate, which we varied in a broad range. Here, we give results for these topologies with UDP traffic over IEEE 802.11 with RTS/CTS enabled and disabled, and for the CXCC protocol in the variants with and without RFA. Since we can adjust the amount of data that is injected in the network freely for UDP, this allows us to draw conclusions on the performance that some *arbitrary* 802.11-based protocol would achieve, if it chose to adjust its output to a certain rate. The comparison with CXCC then shows how the rather different way of packet forwarding with implicit congestion control deals with the same data rate being generated by the application.

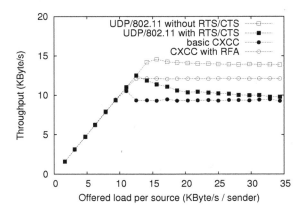

Figure 2.8: Throughput in unidirectional chain.

Figures 2.8–2.10 show how for each of these four protocol variants the obtained through-put develops with increasing source data rate. It can be seen that the UDP traffic is able to sustain good throughput for the chain topology with only one data source. There, 802.11-like packet forwarding seems to be self-regulating. However, as discussed before, for bidirectional traffic the throughput suffers substantially if the input data rate exceeds a very small range. This is visible in both the bidirectional chain and the bidirectional cross topologies. It implies that a congestion control approach on top of 802.11 needs to adjust its output rate quite exactly.

CXCC with the RFA extension is able to achieve and maintain good throughput in all three scenarios. While basic CXCC without RFA suffers from a large number of spurious packet retransmissions, the RFA mechanism allows the protocol to use network resources well and provides effective rate feedback to the sources. By comparing the number of packets delivered for each single connection in the topologies with more than one sender-receiver-pair we found that at least in these simple, symmetric topologies all protocols considered here share the bandwidth fair among the flows.

The optima of the performance of UDP over 802.11 also mark the optimal throughput that can be achieved by any protocol using the common protocol stack. An ideal protocol would be able to find the optimal sending rate for UDP without inducing additional traffic. Thus, it can be seen that CXCC with RFA performs very well in comparison to *any* possible 802.11-based protocol, at least in those simple scenarios, by achieving a throughput close to the optimum and maintaining it for any higher source rate.

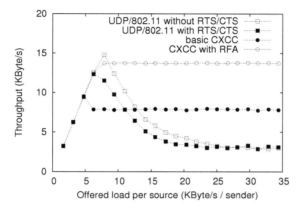

Figure 2.9: Throughput in bidirectional chain.

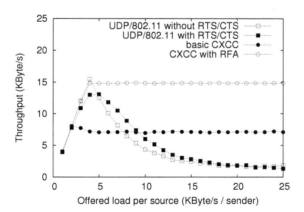

Figure 2.10: Throughput in bidirectional cross.

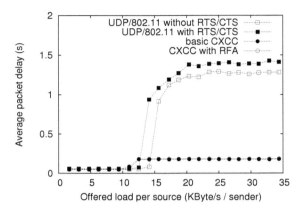

Figure 2.11: Packet latency in unidirectional chain.

The good performance of CXCC is further confirmed by other metrics obtained from the same simulations. We show only the results for the chain and the bidirectional chain topologies here, since the bidirectional cross is generally very similar to the bidirectional chain.

Figures 2.11 and 2.12 show the average packet latency for these two topologies. We define it as the time between the start of the MAC layer transmission of the packet at its source node and the completion of its reception at the destination node. It can be seen that, for UDP over 802.11, the performance deteriorates rapidly with regard to packet latency once the sender's rate exceeds the optimum. The degradation is eminent in all four topologies. In the unidirectional topology, CXCC's packet delay with and without RFA are almost identical. The very good results—in the bidirectional case especially with RFA—again demonstrate its stabilizing properties.

A last evaluation of the performance in the deterministic topology simulations deals with the induced overhead of the protocol. Here we use an overhead metric that quantifies the average amount of data transmitted on the wireless medium in order to bring one byte of payload one hop further. It is computed by summing up the bytes from all packets transmitted on the MAC layer, divided by the product of the amount of application data successfully delivered and the hop distance between source and destination. Note that all transmissions are included, including control traffic and retransmissions of data packets. A protocol without any control overhead, headers, and retransmissions would result in a value of one here. Since wireless communication, especially transmitting, is

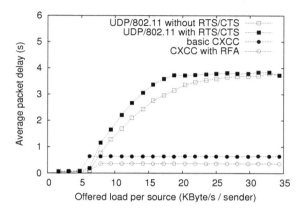

Figure 2.12: Packet latency in bidirectional chain.

rather expensive in terms of energy consumption, this metric also reflects the energy efficiency of the protocol.

Figures 2.13 and 2.14 show the results of this evaluation. It can be seen that again CXCC with RFA outperforms standard 802.11 packet forwarding with UDP largely. The reason is that it ensures that only packets may enter the network which will be able to reach their destination. It therefore does not waste resources on packets that are dropped later on. In the unidirectional topology, a higher number of unnecessary payload data retransmissions heavily increases basic CXCC's overhead. In the variant with RFA there are no unnecessary retransmissions of complete data packets, but at most small RFA packets. This yields an extremely low overhead in comparison to any of the simulated alternatives, which underlines the appropriateness of our design to avoid unnecessary transmissions.

2.7.2 Random Topologies with Long Connections

In the next set of simulations, we examine the steady-state throughput of CXCC in comparison to the three TCP variants in more realistic topologies. The simulated networks cover an area of 1500×1500 square meters, where 150 nodes are placed randomly. Five random connections are set up in each scenario. They continuously try to deliver as much data as possible. The same scenarios are simulated with all considered protocols. Since the large benefits of the RFA extension to CXCC have already become clear, we focus on CXCC with RFA from now on.

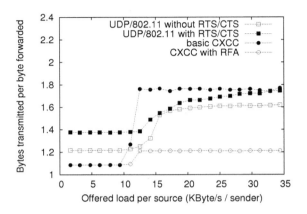

Figure 2.13: Overhead in unidirectional chain.

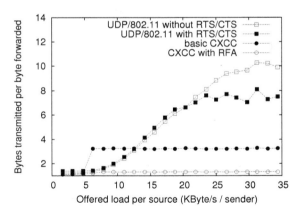

Figure 2.14: Overhead in bidirectional chain.

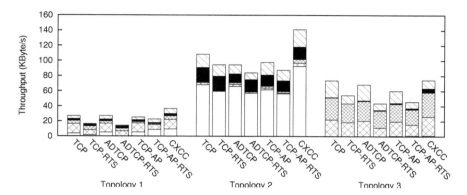

Figure 2.15: Throughput in random topologies with five long-lasting streams.

The throughput of each connection with each of the protocols is measured after an equilibrium has been reached. Figure 2.15 shows the results of these measurements. In the charts, each segment of a bar stands for the throughput of one stream. Within each topology, an identical fill pattern indicates that the respective segment belongs to the same pair of communicating nodes. The chosen representation thus allows not only a comparison of the total throughput, but also of its distribution to the five streams.

As can be seen, some connections are starved completely or almost completely by TCP Newreno. This problem is aggravated further by RTS/CTS. Severe fairness problems with TCP have been reported many times in the literature, and have been traced back to medium capture problems. A primary motivation of TCP modifications and alternatives for wireless multihop networks has always been fairness improvement. While ADTCP and in particular TCP-AP are able to improve the fairness, this comes at the cost of throughput. This is interrelated, because TCP often starves in particular those flows that traverse many hops. A higher throughput for these long flows comes at a high cost, since more medium capacity is necessary to deliver a packet, compared to a flow with few hops. CXCC exhibits better fairness than TCP, without a loss in throughput.

A look at the packet latencies exposes further interesting aspects. In Figure 2.16, we show the average per-hop latency of the packets, i. e., the time from the packet leaving the source node until its arrival at the destination, divided by the hop distance. It is evident that CXCC as well as the TCP modifications achieve a, sometimes very significantly, lower packet latency. While for the TCP variants this comes, as seen before, at the cost of throughput, CXCC combines both low latency and high throughput. The

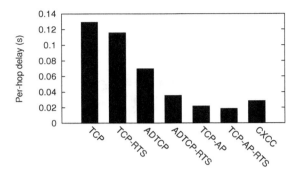

Figure 2.16: Per-hop packet latency in random topologies with five long-lasting connections.

main reasons for the low latency of CXCC are the—by design—extremely short queues, and the resulting short queuing delays. Interestingly, while RTS/CTS generally seems to have a negative impact on throughput and fairness, it can considerably decrease the latency for TCP. This might stem from the lower media utilization with RTS/CTS, resulting in a lower contention level.

An evaluation of the protocol overhead should also take the differences in terms of hop count into account. We calculate the number of transmitted bytes on the medium per payload byte and hop, i.e., how many bytes need on average to be transmitted on the medium in order to bring one byte of payload one hop further. Figure 2.17 shows the results of this evaluation. The significantly higher overhead of all TCP variants results from a higher number of retransmissions. CXCC's RFA mechanism reduces the number of retransmissions of packets with payload to the absolutely necessary minimum.

The hop-count weighted approach also leads to a different throughput measure: TCP's unfairness mainly comes at the cost of streams with a higher number of hops. But, as already mentioned, a high throughput for a stream with few hops can be achieved at a lower media utilization. Thus, in Figure 2.18, we provide the same results as before in Figure 2.15, but with each connection's throughput weighted by the number of hops along its route. It shows that CXCC is in fact able to utilize the network best, since long connections gain more throughput and thus the per-hop throughput is higher.

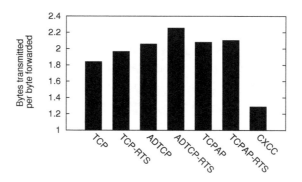

Figure 2.17: Overhead in random topologies with five long-lasting connections.

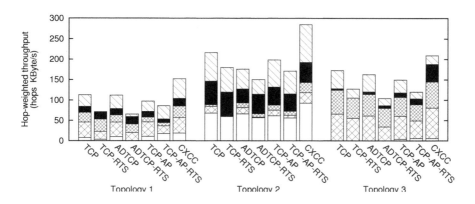

Figure 2.18: Throughput in random topologies with five long-lasting streams, weighted by their hop count.

2.7.3 Random Topologies with Dynamic Traffic Patterns

Through the last type of simulation that we have performed we examine how well CXCC behaves in a more realistic and dynamic scenario. We consider random topologies of the same dimensions and node counts as above. Now, 120 short data transmissions are scheduled between random pairs of nodes, each starting at a random time between 0 and 120 simulation seconds. Each of these transmissions has a random, uniformly distributed amount of data in the range between 5 and 50 kilobytes to deliver.

Again, we have performed all simulations of all TCP variants with and without RTS/CTS enabled. Like for most aspects before, the performance with RTS/CTS enabled is generally worse. Therefore, for space and readability reasons, we show only the results without RTS/CTS here.

Figure 2.19 depicts the packet reception times of some randomly selected streams in one of the simulation runs for the two-ray ground model and for the shadowing model respectively. Each point denotes one packet reception. The y-coordinate of each point denotes the stream that it belongs to, while on the x-axis the packet arrival time is shown. The dashed horizontal bars show, for each single connection, the time span from when the connection starts to the delivery of the last data segment. All four sub-figures are based on the same streams from the same scenario, and thus the node positions and communication partners are identical, equal amounts of data are to be transmitted and the transmissions start at the same times. To avoid misinterpretations, only the first successful reception of a segment by the receiver is shown upon duplicate deliveries.

It is visible that, for the vast majority of transmissions, CXCC not only delivers the last segment much earlier than all TCP variants, but it is also able to sustain a smoother rate. We attribute this to a faster, since implicit and thus practically immediate, adjustment to the optimal rate in case of changing network conditions. Our interpretation of the bad results for all TCP versions is that the feedback path is too long to yield accurate information on the network state if the traffic pattern is as dynamic as in these simulations. During some of the transmissions, multiple losses of data segments and acknowledgments and the resulting long retransmission timeouts cause long periods of inactivity. Therefore, a number of transmissions that in fact start quite early are not able to complete for a long time, even when the network is idle.

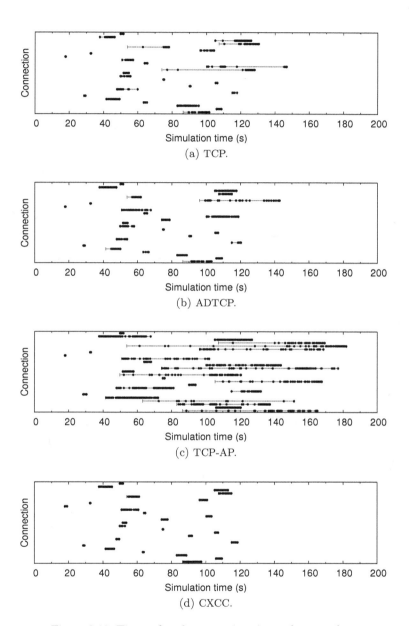

Figure 2.19: Times of packet receptions in random topology.

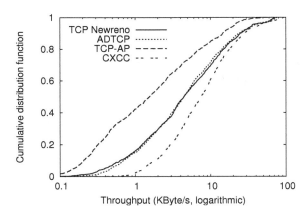

Figure 2.20: Cumulative distribution functions of stream throughputs in random topologies with dynamic traffic patterns.

Finally, we now consider the per-stream throughput and the fairness of the different protocols under rapidly changing traffic patterns. To examine this aspect, we performed more simulations in different random topologies, all with their key parameters chosen as described above. Figure 2.20 shows the cumulative distribution functions of the throughputs of all the streams. Note the logarithmic x-axis. We calculate the throughput by dividing the amount of data by the transmission duration. The transmission duration is defined as the time between the start of the transmission, when the packets are enqueued at the source node, and the point in time when each segment has been received at least once by the destination node.

Apart from a general trend to higher throughputs, a significantly better fairness of CXCC is evident: while for the TCP variants many streams have a low or very low throughput, the CXCC connections all obtain at least some minimum share of the bandwidth. This can be seen from CXCC's curve, which starts increasing comparatively late, implying that very few connection achieve a low or very low throughput. The key reason for the better fairness of CXCC is that the nodes refrain from capturing the medium for a long time. After a packet transmission a node is forced to stop sending more data from the same stream. It has to wait for the reception of an implicit or explicit acknowledgment, or until the retransmission timeout elapses. This gives other nodes the opportunity to start or continue transmissions.

2.8 Real-World Testbed Results

In the previous section we have presented simulation results, indicating that our implicit approach of performing congestion control is in fact able to provide an efficient way of protecting the network from overload. However, simulations are not able to model all factors that might influence a protocol in the real world. Therefore, we complement them with some measurement results from an implementation of CXCC with RFA in a real hardware testbed.

As stated before, CXCC cannot be implemented on today's commodity 802.11 wireless hardware. There, a large part of the MAC functionality is realized very close to the hardware, in the (proprietary) firmware, which is not accessible for modifications. When looking for a way to overcome these difficulties we came across the Embedded Sensor Boards (ESB). These relatively inexpensive devices were developed at the Freie Universität Berlin as part of the ScatterWeb project [Fre]. They are intended to serve as a testbed platform for wireless sensor networks. ESB nodes are battery-powered and equipped with a collection of sensors and a wireless interface. For our purposes, however, their main advantage is the open firmware, which allows modifications to every part of the software, down to the manipulation of each single bit transmitted on the wireless medium. Here, they are used as devices in a wireless multihop testbed instead of their original purpose of being used in sensor networks.

Of course the non-802.11-compatible physical layer of the ESB nodes, operating at only 19.2 KBit/s in the 868 MHz band, does not allow for a direct performance comparison to 802.11-based networks. But our main intention is to show that the concepts of CXCC work in practice and exhibit a behavior similar to that in the simulations.

Since on the ESB nodes there is not as much software "infrastructure" available as it can taken for granted on, e. g., PDAs or a PC, it was not sufficient to implement only the CXCC protocol. We also created the rest of the necessary testbed infrastructure. This comprises, for example, a logging facility that is able to log a large enough number of MAC layer events in the limited storage space available (64 KB in each node), a static routing module, and traffic generators. Additionally, some convenience tools for routing table generation and distribution, for the verification of the topology and for the collection of log data have been created. More details on this experimental framework can be found in [JSLM06b].

In addition it was necessary to make modifications to the standard ESB firmware's MAC and link layer to resemble 802.11 more closely. For example, the number of packet

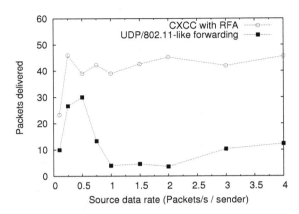

Figure 2.21: Measured throughput in bidirectional chain topology experiments.

retransmission attempts has been set to seven for those experiments where CXCC was not used. Also the ESB nodes are—due to their very limited hardware resources—not able to handle packets of the size that is common for IEEE 802.11. Since there are only 2 KB of RAM available in total, there is not enough space for queues containing long packets. In order to be able to use a reasonable data packet size, we prepended additional 200 bytes to the data packets, as some kind of additional preamble. We used 32 bytes of payload per packet in our experiments, a size that is easily feasible with the ESB nodes. With the prepended 200 bytes, a 32 byte data packet transmission results in the same medium occupancy as a "real" 232 byte packet would. Control packets such as ACKs or RFA packets are transmitted without the artificially increased preamble, so they occupy the medium with their regular size.

In our experiments we used six ESB nodes, set up in a bidirectional chain topology. Each node was placed in a different room. This was sufficient to prevent a reliable direct communication between nodes that are not neighbors. In each of three separate experiment runs we slowly increased the offered load at the nodes for both CXCC with RFA and 802.11-like packet queuing and forwarding. At each examined data rate, traffic was generated for two minutes. Then the successfully delivered packets were counted. Figure 2.21 shows the outcome of our throughput measurements, averaged over the three experimental runs.

The results show clearly that the real-world behavior of both protocols matches the simulations very closely. Of course the absolute values are very different—but this is hardly surprising given the vastly different physical layers. However, much more importantly, on a qualitative level, the 802.11-like approach's performance drops after some optimal input rate is exceeded, while CXCC's throughput remains stable at a comparably high level. We take this as a confirmation that the throughput-stabilizing properties of CXCC are also present in real networks.

In the course of conducting and analyzing the experiments and debugging the protocol implementation we have encountered a central problem of experimental research in wireless multihop networking. The analysis of events in the network nodes and their direct and indirect effects depends on the log files written by the network nodes during the experiment. The clocks in these nodes are, however, typically not very accurate. Therefore, the event timestamps in the log files deviate—a fact that heavily impedes the reconstruction of event sequences and the identification of temporal correlations. Later in this thesis, in Chapter 6, we will introduce a novel way to overcome this problem.

2.9 Chapter Summary

We have proposed a novel way of accomplishing congestion control in wireless multihop networks: implicit hop-by-hop congestion control. It is based on the insight that an input rate exceeding the optimal output rate of a node or network area even on a short-term will be detrimental for the performance of a wireless multihop network. Our mechanism exploits the wireless broadcast medium in order to gain the necessary information for a backpressure mechanism that reliably limits the number of packets to one per flow and hop, and thereby implicitly avoids network congestion. We have presented a protocol, CXCC, that builds upon the idea of implicit hop-by-hop congestion control. An improvement of the CXCC protocol, the Request For Acknowledgment (RFA) mechanism, avoids unnecessary data packet retransmissions.

Our simulation results demonstrate that in simple and deterministic scenarios as well as in more realistic, random ones CXCC is able to effectively adjust the packet sources' rates and to utilize the network capacity well. In comparison to TCP and two other transport protocols for mobile ad-hoc networks, good fairness properties and a very competitive throughput can be observed. All this shows that implicit hop-by-hop congestion control as a new congestion control paradigm not just works well, but also exhibits some remarkable advantages over common transport layer end-to-end mechanisms. In particular these are the ability deal with UDP- as well as with TCP-like traffic, very fast reaction times, a low packet delay, very good energy efficiency, and a simple protocol design. The latter greatly eases the adaption of the protocol to new usage scenarios and environments. Since wireless multihop network applications span a broad range of traffic types, we consider this a very central benefit.

The simulation results are accompanied by measurements from an implementation on real hardware. We have shown that the behavior of CXCC in a real network matches the expectations from the simulations. The chosen hardware platform allowed us to do such an implementation by avoiding the constraints imposed by commodity 802.11 hardware.

Chapter 3

Implicit Reliability: BarRel

In the previous chapter, we have introduced a novel way to perform congestion control in wireless multihop networks, as an alternative to TCP-like observation of and inference from packet losses. One major benefit of the devised protocol, CXCC, is that it does not require explicit feedback messages to the source, neither from the destination nor from within the network. In particular, this also makes end-to-end acknowledgments unnecessary for congestion control purposes.

But in TCP—and likewise in many other transport protocols—these acknowledgments serve a dual purpose: they are not only used to obtain congestion feedback, but also for end-to-end reliability provisioning. It seems to be a common assumption that such acknowledgments are indispensable for end-to-end reliable unicast communication with TCP service semantics [Git76]. But it has been observed that these acknowledgments are a major source of performance degradation in wireless multihop networks. ACK packets typically travel through the same network regions as the data packets, in reverse direction. Since all transmissions within the same area share the medium, the ACKs cause significant additional network load and increase medium contention. Therefore, end-to-end acknowledgments are often considered a problem in wireless multihop networks [dOB07, AJ03, dMCDA02].

In the context of our work, the question arises whether not only congestion control, but also end-to-end reliability can be realized in an implicit way, i.e., without explicit reliability feedback from the destination back to the source. And indeed it turns out that this is possible. Our approach heavily relies on the characteristics of CXCC and on the ability of the routing protocol to accurately determine the distance towards a destination. We do thus not claim that the approach shown here may be applied to all networks. Rather, our core contribution is to demonstrate that end-to-end reliability *can* be ensured without using end-to-end acknowledgments, and that such an approach can exhibit significant benefits.

The core idea of the *Backpressure Reliability* (BarRel) transport protocol presented here is as follows. Assume that a congestion control approach limits the number of packets that are underway in parallel in the network by a function f of the route length n, and that this route length is known. Both prerequisites can be fulfilled if CXCC is used for congestion control and AODV [PR99] is employed as a routing protocol. The ability to inject a packet into the network may then provide information about the reception of preceding packets. For example, if it is possible to transmit packet $f(n) + 1$ then one of the $f(n)$ preceding packets must have reached the destination—otherwise the congestion control approach would not have allowed the transmission. Of course there are some subtleties involved, in particular regarding route failures. We will demonstrate that it is nevertheless possible to build a practically usable protocol based on this key idea.

In the following discussion of BarRel, we first describe the basic variant of the protocol to convey the general idea. It uses one explicit end-to-end acknowledgment at the end of each (arbitrarily long) packet burst. We then present a modified variant that allows for reliable transmission without any oncoming multihop traffic *at all*. These protocol variants, their design, and their properties are also the topic of a paper that is currently under review [SKLM].

3.1 Related Work

Congestion control and reliability are traditionally tightly intertwined issues, so most proposals deal with both at once. A general survey of transport layer concepts, techniques, and protocols can be found in [IAC99]. Many approaches that are related to implicit hop-by-hop congestion control and CXCC have already been discussed in Section 2.1. In the following, we concentrate on the reliability aspect over a wireless multihop medium: how do other protocols achieve end-to-end reliability?

To alleviate the problem of acknowledgment traffic in MANETs, it has been proposed to reduce the number of acknowledgments, either by using the standardized TCP delayed ACK (DACK) mechanism [Bra89] as it has been done, e. g., in [SM01, DB01], or through improved variants of it, like those by de Oliveira and Braun [dOB07] and by Altman and Jiménez [AJ03]. All these approaches, however, reduce the number of acknowledgments at most by some small, constant factor—standard DACK combines up to two acknowledgments into one, [dOB07] and [AJ03] up to four. For continuous data transmission, a continuous stream of acknowledgments remains necessary.

In their Contention-based Path Selection (COPAS) scheme, Cordeiro et al. [dMCDA02] use different paths for forward and reverse traffic, to reduce intra-flow contention between data and ACK packets. It remains, however, inevitable that forward and reverse traffic share the medium at least around sender and receiver. Moreover, the probability of a route break increases if different routes are used, because both forward and reverse path need to stay intact for the data flow not to be interrupted.

Split TCP by Kopparty et al. [KKFT02] uses TCP as a basis, but separates congestion control and reliability. It establishes so-called proxies in intermediate nodes along the route. The proxies buffer packets and transmit them either to the next proxy or to the final destination over a small number of hops. The standard TCP acknowledgment scheme is employed between the proxies. In addition to these acknowledgments, end-to-end ACKs are used to detect and overcome possible proxy failures. Thus, in total, Split TCP even increases the amount of acknowledgment traffic.

The MANET-specific transport protocols ATP [SAHS03] and EXACT [CNV04] both rely on the nodes within the network to provide piggybacked feedback on the congestion state, which is then transmitted back to the senders. So, these protocols still require continuous end-to-end feedback for both congestion control and reliability provisioning. OPET [ZF06] is used with TCP on top for reliable service. Therefore, this approach does also not avoid end-to-end ACK traffic.

The Transport Protocol for Ad-hoc Networks (TPA) by Anastasi et al. [AACP05] is inspired by TCP, but the protocol is designed to minimize the number of end-to-end packet retransmissions. TPA also uses end-to-end acknowledgments. Packets are grouped into blocks, and no data from the following block is allowed to be transmitted before all packets from a block are acknowledged.

In wireless sensor networks, reliable unicast data streams are not the predominant traffic pattern. So, sensor network transport mechanisms are typically not general-purpose, transparent TCP replacements. Where, like with Distributed TCP Caching (DTC) [DAVR04], fully TCP-equivalent service is provided, acknowledgment packets are used.

3.2 The BarRel Transport Protocol

The key idea of BarRel is founded on properties of the network that arise if a backpressure mechanism like that of CXCC is used: such a network allows a maximum of one (untransmitted) packet to be queued in each node along the route. This establishes an upper bound on the number of sent packets that have not yet arrived at the destination: for a connection that runs over n hops, at most n packets may be underway in the network at any given point in time. The tightly limited queue length also implies that packets are never lost in the network unless simultaneously a route break occurs: there are no queue overflows, so packets are only dropped if at the same time a broken link is detected.

Let us, just for the moment and for the purpose of discussion, disregard the existence of route breaks. Under these constraints, the fact that further packets may be sent implies that earlier packets must have successfully arrived. In a network without route breaks, explicit end-to-end ACKs are thus unnecessary for a continuous data flow: being allowed to send out the i-th packet by itself constitutes an acknowledgment for packet $i - n$. We formalize and prove this with the following theorem.

Theorem 3.1. *Let* $N_0, \ldots, N_n$ *be the nodes on the n-hop route from source N_0 to destination N_n, and let p_i denote the i-th packet. When N_0 is allowed to send p_i, $i > n$, then N_n has successfully received and (single-hop) acknowledged p_{i-n}.*

Proof This can be shown by a simple induction proof. The one-hop case, $n = 1$, is trivial.

For the induction step $n \to n + 1$, the route consists of the nodes $N_0, \ldots, N_{n+1}$. If N_0 sends p_i, N_n has acknowledged p_{i-n} by the induction hypothesis. But N_n may only have sent (and thereby acknowledged) p_{i-n} if N_{n+1} has acknowledged $p_{i-(n+1)}$. The assertion follows. $\qquad\square$

Knowledge of n is obviously necessary to exploit this property. Most topology-aware routing approaches, however, can supply this value easily. For the purposes of discussion and evaluation, we will use the AODV routing protocol [PR99].

3.2.1 Node and Link Failures

Since the only possible reason of packet losses are broken links, a packet loss implies a route break. As long as the route remains stable and all intermediate nodes can eventually forward the data packets, the basic BarRel mechanism as discussed so far hence allows for reliable, completely ACK-free data transport. The sender will, n packet transmissions later, implicitly learn about the safe delivery of a segment. But especially in a mobile environment, routes will not remain stable over arbitrary long timespans, and intermediate nodes are not unlikely to fail during the lifetime of a connection. The protocol needs to be able to deal with these circumstances.

When a route change occurs, the source can not know for sure whether the packets that have been in the network at that time have been delivered or not. However, at most n packets can possibly be in the network at once. After a route change, the source may thus simply repeat the last n packets. By appropriate interaction with the routing protocol, we ensure that *i)* a source node eventually learns about every route change, including information on a possibly different new route length n, and *ii)* until this notification has arrived, packet forwarding along the old route does not proceed, i. e., no further implicit or explicit ACKs may occur. Then, the number of packets that may be lost never exceeds one per node, and their total number is limited by the (old) route length. Combined with the backpressure reliability mechanism discussed so far this is sufficient to ensure full end-to-end reliability.

While similar solutions seem possible for other routing protocols we discuss in the following how these two criteria can be guaranteed for one specific routing protocol, AODV. With standard AODV, they are not always met. For example, it may happen that two connections share parts of a route, and a broken route is repaired by one of them— potentially resulting in a different route length, and without the other connection being aware of that fact.

A small extension to standard AODV is sufficient to guarantee the notification of each source about every route change, and to maintain backpressure also in case of route errors, thus satisfying both criteria. We build upon the destination sequence numbers in the AODV routing tables. Observe that routing entries are only updated if either a new route with a higher destination sequence number is discovered, or one with the same sequence number, but a lower hop count. Hence, as long as both destination sequence number and hop count remain stable in all nodes along the route, the route itself also does not change. We exploit this trait, by ensuring that the pair of destination sequence number and hop count distance is consistent along the whole route. In the end, this is nothing else than making sure that the source node's view of the route is up-to-date.

We piggyback the source node's current values of destination sequence number and hop distance onto each packet. The intermediate nodes may then check whether their current values are consistent with the piggybacked ones in an incoming packet. If not, the routing between this node and the destination has possibly changed. In that case, the forwarding is rejected; instead the upstream nodes' routing tables are updated. For the latter purpose, a *Sequence Number Notification* (SNN) packet is sent. It contains the up-to-date sequence number and hop distance, which are thereby propagated back towards the source. In a certain sense, an SNN can be seen as a combination of route error notification and follow-up route reply: it invalidates the previously existing routing entry, but at the same time establishes a new route. An SNN that arrives at the source node indicates a changed route and thus triggers a retransmission of the last n packets.

If, after a route failure, no routing entry exists in an intermediate node that receives a packet, a route error message must have been lost. The packet reception is thus answered with a respective notification. In essence, this "revives" lost route error messages, when upstream nodes continue to use the route.

Since the forwarding is rejected if the routing information in the packet is not up-to-date, no further forwarding and thus no implicit ACKs will happen until the information arrives at the source. Backpressure is thus maintained. As a result, the proposed mechanism can guarantee that the sufficient criteria for notifying the sources about route changes are met.

In most cases, SNNs are not necessary. Typically, a source node will already learn about route changes by receiving a route error message, and subsequently initiate a new route discovery. But SNNs guarantee that the vital semantics for BarRel are also maintained in all other cases—including, for instance, a lost route error notification and the use of local repair mechanisms. One might argue that propagating this information back towards the source and repeating n packets whenever packets *might* not have been delivered is not always the most efficient option. While this is surely true, this procedure allows to keep the complexity of the mechanism very low. Moreover, the performance figures presented later on strongly underline that potential small losses in the rare cases where SNNs and unnecessary retransmissions actually occur are more than compensated by the huge gains obtained while the routes are stable.

3.2.2 Sequence Numbers and Order Preservation

Due to retransmissions, multiple copies of some packets might arrive at the destination. These duplicates need to be identified and filtered. To accomplish this a sequence number is assigned to each segment. They closely resemble TCP sequence numbers, in that they are kept constant for retransmissions. They can also be used for order preservation at the receiver. In this context, there is an interesting property of the protocol, which largely eases the processing in the destination node. When a segment arrives at the receiver, all previous segments must have arrived, too. "Gaps" caused by missing packets or packet reordering cannot occur. Thus, a simple duplicate suppression mechanism is sufficient, packet ordering is then automatically preserved.

3.2.3 Acknowledging the Last Packets: TACKs and TRFAs

So far, we have only considered continuously sending sources. In that case, BarRel, as described up to this point, achieves reliable data delivery without end-to-end acknowledgments. Let us now look at the situation where less than n packets remain to be transmitted, either temporarily at the end of a packet burst, or at the end of the connection. How can the reliable delivery of these last n packets be accomplished?

We propose two mechanisms to deal with the last packets of a burst. The first one is rather straightforward: if some packet is the last one in a burst of packets generated by the source, the sender requests by a flag in the packet header (the *LAST flag*) that an explicit end-to-end *transport layer ACK* (TACK) shall be sent.

When the source sets the LAST flag, it schedules a timer. If no corresponding TACK has arrived when this timer expires, two situations are possible: either everything is fine, but the TACK is delayed, or a route failure has prevented either some of the data packets or the TACK from getting through. To deal with a TACK timeout, the source generates a *transport layer RFA* (TRFA) packet. It is an end-to-end request, intended to find out about the fate of the packets. Indeed it is quite similar to CXCC's single-hop RFAs. TRFAs are subject to the congestion control's backpressure. Note that therefore, when the destination receives a TRFA, it must also have received the corresponding last data packet. A node thus always responds to a TRFA by repeating the respective TACK.

Sending out a short TRFA that requests for an end-to-end acknowledgment is *always* superior to retransmitting the much larger data packet. This is because the TRFA can never possibly arrive if the previous data transmission did not arrive. If the last data packet has *not* arrived, this can only be due to a route failure. Then the TRFA will encounter the same problem, and, if necessary, trigger an SNN; in some sense, the central purpose of the TRFA is to "check" the route for route failures, where the notification did for some reason not make it to the source node.

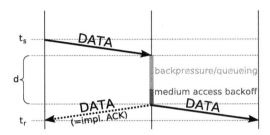

Figure 3.1: Measuring the forwarding delay.

To set the TACK timeout appropriately, we use an adaption of Jacobson's variance supported retransmission timeout (RTO) algorithm for TCP [Jac88]. If TACKs for the connection have already occurred recently, they can of course be used as samples. This will, however, not always be the case. So, we need to obtain an estimate in a different way. We do so by observing the forwarding delay of the first forwarder. This forwarding delay, denoted by d, is the time between receiving the packet and starting the transmission forwarding it further on, as it is shown in Figure 3.1. It includes the delay due to queuing and backpressure, and possibly also medium access backoff times. d can be observed by the source node of the stream: the times t_s when the packet transmission at the source node starts and t_r when the implicit ACK is overheard are known, as well as the packet size and the used medium bandwidth.

The delay at the first forwarder also conveys information about the situation further downstream: it may not forward packets at a higher frequency than its successors. It may well be that nodes further downstream can transmit packets more quickly—consider a route first leading through a tight bottleneck in a network region with heavy load, and then through an area of low network traffic, where packets flow freely. Due to the backpressure mechanism, the opposite situation is, however, not possible once an equilibrium has been reached and the route is "filled" with packets. So d as measured at the first hop is an upper bound estimate for the forwarding delay at any node along the route. If n is the route length in hops, $n-1$ forwarding delays will occur during the transport of the data packet to the destination node, and n forwarding delays for delivering the acknowledgment back to the source. We can thus obtain an estimate for an upper bound of the total forwarding delay:

$$D = (2n - 1) \cdot d. \tag{3.1}$$

We use these forwarding delay samples as the input for Jacobson's algorithm, the time-out is based on its output. Note that D and consequently also the output of the RTO algorithm do not include any transmission times, so they are independent from the packet size. We add the transmission times for n data packet and n TACK transmissions to obtain what we use as our TACK timeout.

We want to stress here that the timeouts are much less critical for BarRel than they are for TCP. In BarRel, timeouts are used only at the end of packet bursts, and they do not trigger retransmissions of data segments, but only (small) TRFA packets. Thus, in particular, a too optimistic round trip time estimate will have only very limited negative impact, since it does not, as for TCP, cause many unnecessary data packet transmissions.

3.2.4 CaRe Packets

The TACK/TRFA mechanism has the important advantage of a very low control packet overhead, and it eliminates almost all oncoming ACK traffic: TACKs are only necessary at the end of transmission bursts. Although the problem of acknowledging the last packet is solved, the mechanism obviously stands in sharp contrast to our otherwise consistently implicit approach. Besides that, choosing an appropriate TACK timeout involves much effort. Therefore, it is worth thinking about alternatives.

Let us turn back and look again at the initial point of our considerations. TACKs were introduced because there were no data packets left to confirm the delivery. But instead of requesting explicit end-to-end feedback, a source may also just generate the requested n additional packets: it may send n empty placeholder packets succeeding the last data segment, in some sense "refilling" the send buffer of the source node according to the route's "packet capacity". If all these *Capacity Refill* (CaRe) packets have left the source node, all preceding data packets must have successfully arrived.

As a further optimization, CaRe packets do not need to travel the entire n hops to witness the forwarding of the preceding packets. They may be dropped once it is sure that the last data packet has reached the destination. This can easily be accomplished by limiting the TTL of the j-th CaRe packet to $n - j + 1$ hops.

There are many positive features of the CaRe approach. First of all, it is far less complex than the timeout-based TRFA mechanism. It makes use of the backpressure concept of BarRel and CXCC, avoids any oncoming traffic, and therefore seems well-suited for networks with a shared broadcast medium. Moreover, it also guarantees a very fast notification about successful arrival or loss, since no phases of passive waiting for an ACK or a timeout are required.

On the other hand, however, the amount of control packets is higher for CaRe packets. For TRFA/TACK, in the best case, acknowledging the end of the packet burst by a TACK requires n single-hop control packet transmissions, to deliver the TACK triggered by the LAST flag. Up to n TRFAs may have to be transmitted—so the worst case effort for the TRFA/TACK scheme is $O(n^2)$ single-hop control packet transmissions. For CaRe packets, the effort is *always* quadratic in the route length, not only in the worst case. Thus, CaRe packets may be a bad choice if many short data packet bursts are sent and the route is long. But note that it is possible to use both TRFAs and CaRe packets in the same network, or even within the same connection: at the end of each burst, a sender may decide to use either the LAST flag and TRFAs, or to send CaRe packets, based on, e. g., the application's preferences, observations of the traffic pattern, or the route length.

3.2.5 Other Transport Layer Functions

For a complete transport protocol, not only congestion control and reliability provisioning are necessary. Order preservation has already been discussed above; flow control, a way to establish and close connections, and port addressing are also provided by BarRel. It is therefore possible to implement BarRel with a TCP-compatible socket interface, hence not necessitating any changes to existing applications.

For short-term flow control, if the buffer at the destination node runs full, it will send KALs instead of explicit CXCC ACKs. Consequently, backpressure will build up, preventing the source node from injecting more data packets into the network.

BarRel adopts TCP's mechanisms for connection establishment and termination. To set up a connection, a TCP-like SYN–SYNACK–ACK handshake is used, where, of course, the ACKs can again be implicitly obtained. Likewise, to close a connection, a FIN-flag in the BarRel header is used, where the delivery of the packet with the FIN flag set may be acknowledged by one of the mechanisms discussed above.

3.3 Evaluation

To examine the performance of the proposed protocol, we have performed simulations with the ns-2 network simulator [ns2a]. We focus on the aspects that are particularly relevant with regard to the presented reliability mechanisms. We thus concentrate on dynamic and mobile scenarios, where route breaks and corresponding packet losses are common. In more simple, static scenarios with few, long-lived flows, BarRel does not add significant additional overhead to CXCC. Then, the performance is dominated by CXCC's congestion control mechanism. Detailed examinations of CXCC's performance can be found in the previous chapter and in [SLM08].

Here, we compare the BarRel/CXCC protocol stack to standard TCP Newreno [FHG04], TCP-AP [EKL05], and ADTCP [FGML02] over IEEE 802.11.

3.3.1 Methodology

We use packets with 512 bytes of payload. 200 nodes move on a square area of $1500 \times 1500\,\mathrm{m}$ according to the random waypoint (RWP) mobility model. It is the most widely used model in the context of this work. To overcome its well-known limitations [YLN03], we use a minimum node speed of 10 % of the maximum speed in the respective simulation, and choose the initial positions and speeds according to the stationary distribution of the RWP model as described in [NC04]. We do not use pause times. As mentioned above, AODV [PR99] is used for wireless multihop routing. The results presented here have again been obtained with the RTS/CTS mechanism of 802.11 disabled, for the same reasons as discussed before.

A problem with the evaluation of random topology simulations with many connections is that both throughput and packet latency are not directly comparable between connections over largely different hop count distances and in different scenarios. Therefore the arithmetic mean of, for instance, the throughputs of all connections in the simulated topologies is not a suitable performance metric. In particular, it overemphasizes connections with high throughputs. The resources that need to be spent in order to deliver one packet vastly differ between connections of different hop counts. Thus, the throughput of a connection over a very short distance can be increased largely if connections over more hops are assigned a smaller share of the bandwidth. Consider a situation where one connection with high throughput increases its throughput again by, say, 10 %, but this comes at the cost of starving some low-throughput connections almost completely. The arithmetic mean may nevertheless increase.

Normalizing throughput and latency with respect to the hop count is also not appropriate: with dynamic routing, this value may often change during a connection, and because of unforeseeable interactions between transport, routing, and MAC layer it is also not guaranteed that identical routes or even just routes with the same hop count will be used when different transport protocols are simulated.

To overcome this problem at least partially, we use the geometric mean for throughput and packet latencies over connections with different hop counts. This guarantees that, in a certain sense, each connection has the same impact on the overall metric, regardless of its absolute value: if, e.g., the throughput of one connection doubles, while the throughput of some other connection is cut in half, the geometric mean of the throughputs remains unchanged.

We show 95 % confidence interval error bars in our graphs. They are based on the assumptions that the samples are i. i. d. and normal or—for geometric means—lognormal distributed.

3.3.2 FTP Traffic

The first set of scenarios uses FTP-like traffic between a varying number of randomly chosen node pairs. For each connection a pair of nodes and a file size in the range between 5 and 50 KB are randomly chosen. Each connection is initiated at a random point in time between 0 and 120 simulation seconds. Then, the source node transfers a "file" of the chosen size to the destination. The simulations run until all data is successfully delivered. We simulate all considered protocols in the same set of topologies, with identical mobility patterns, node pairs, file sizes, and start times, thus confronting each protocol with exactly the same task.

In Figure 3.2, we keep the maximum node speed fixed at 6 m/s and vary the number of connections between 10 and 70, thereby gradually increasing the network load. We then measure the mean throughput of the connections by dividing their respective file size by the time from initiating the connection at the source node until the successful arrival of the last data segment at the destination.

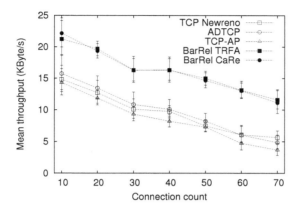

Figure 3.2: Mean FTP throughput with varying network load.

It is evident that the BarRel/CXCC protocol stack substantially outperforms standard TCP as well as the modified TCP variants for MANETs. The mean throughput is substantially higher with BarRel than with any of the TCP variants.

There is no significant difference between BarRel with TRFAs for acknowledging the end of a packet burst and the entirely end-to-end ACK-free variant with CaRe packets. This is not too surprising, since CaRe packets or TACKs and TRFAs both occur only at the end of packet bursts, and there is only one single burst per connection in the FTP simulations. Thus, here, the specifics of these mechanisms can have only very limited impact on the overall performance.

Note that the single absolute values that go into the mean values shown in the plot are vastly different—there are connections with less than 1 KB/s and others in the order of 100 KB/s. We visualize the distribution of the throughputs in Figure 3.3. It shows the cumulative distribution function of the single streams' throughputs for the case of a maximum node speed of 6 m/s and 40 connections (note the logarithmic x-axis). The two BarRel variants on the right hand side are so close together that their lines are hard to distinguish. Their significantly higher throughput is again evident. Connections with low throughput seem to profit most. Consequently, the fairness with BarRel is substantially better than with any of the TCP variants: there are much fewer connections with low or very low throughput.

But not only the throughput with the BarRel variants is substantially higher, other metrics also confirm the positive picture. In Figure 3.4, we analyze the mean packet

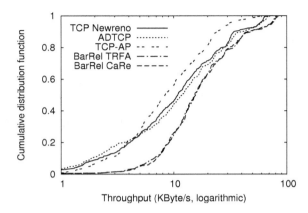

Figure 3.3: Cumulative distribution function of FTP throughputs.

latency, i.e., the time it takes an average packet from leaving the source node until arriving at the destination. The again substantial benefit of BarRel can be traced back to the short queues of the underlying CXCC congestion control mechanism: the low number of packets in intermediate nodes avoids long waiting times in forwarding queues along the route. Only TCP-AP is able to achieve even slightly lower latencies—however, as seen before, at the cost of a lower throughput.

Figures 3.5 and 3.6 show the impact of varying node mobility on the considered protocols. They are the counterparts of Figures 3.2 and 3.4. We keep the number of connections fixed at 40 and instead vary the maximum node speed in the simulations. It can be seen that node mobility has only limited impact on the performance of all considered protocols.

Another central metric for the performance of a wireless multihop communication protocol is its protocol overhead. We measure it as before by looking at the number of bytes transmitted on the wireless medium per byte of delivered payload, including all headers, control messages, and retransmissions. We show the overhead in Figure 3.7. It is evident that TCP as well as all considered TCP variants waste a lot of medium capacity, in particular for retransmissions.

One reason for the good performance of BarRel is that the BarRel/CXCC protocol stack avoids unnecessary re-routing. The soft timing principle in CXCC—allowing feedback to arrive with significant delay—largely reduces the number of route errors, and thereby also the amount of spurious routing traffic in the network. To substantiate this claim,

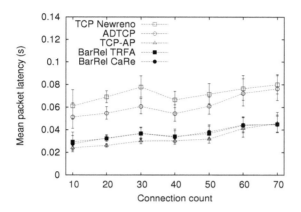

Figure 3.4: Mean FTP packet latency with varying network load.

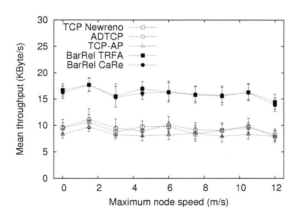

Figure 3.5: Mean FTP throughput with varying node mobility.

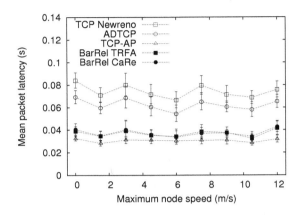

Figure 3.6: Mean FTP packet latency with varying node mobility.

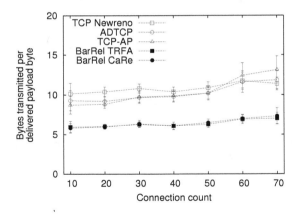

Figure 3.7: Protocol overhead for FTP traffic.

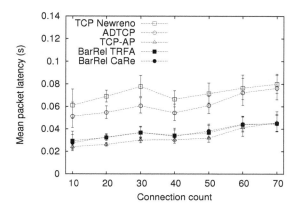

Figure 3.4: Mean FTP packet latency with varying network load.

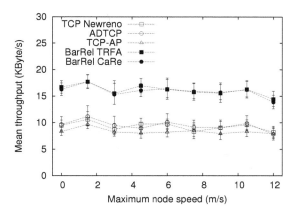

Figure 3.5: Mean FTP throughput with varying node mobility.

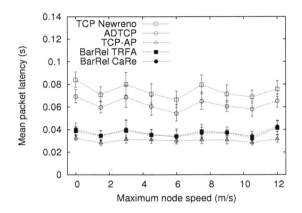

Figure 3.6: Mean FTP packet latency with varying node mobility.

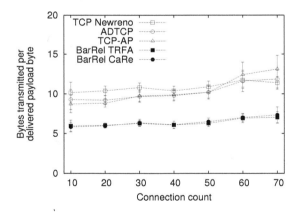

Figure 3.7: Protocol overhead for FTP traffic.

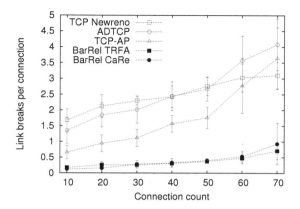

Figure 3.8: Average number of link break callbacks per FTP connection.

we show the number of link break callbacks per stream that occurred in our simulations in Figure 3.8, again for 6 m/s maximum node speed and a varying network load. For an increasing number of connections, their number steeply increases for the TCP/802.11-based protocol stacks, while it is at a much lower level for the BarRel/CXCC protocol stack.

3.3.3 HTTP Traffic

It has already been mentioned that the FTP connections in the simulation results presented so far consist only of one single burst of data. BarRel's reliability mechanisms, however, are particularly critical in the case of many short data transmissions. We therefore complement our simulation results with measurements from a different traffic pattern. In these HTTP traffic simulations, up to five nodes are designated as HTTP servers and each server is assigned three nodes as clients. These clients request "pages" from their respective servers.

We simulate non-pipelined, persistent HTTP 1.1. This yields an interactive traffic pattern where short bursts of data are alternately transferred in both directions between client and server. Each client requests five pages from its server, at randomly chosen points in time, again between 0 and 120 simulation seconds. A page consists of between one and ten "objects", each object has a random size of up to 10 KB. These objects are fetched sequentially, one after the other. As a performance metric, we measure the time from initiating a page request at the client until the last object of the page is

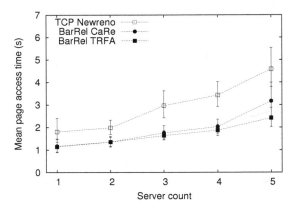

Figure 3.9: Mean HTTP request latency with varying network load.

downloaded completely. Again, we use the geometric mean of these page access times to obtain global performance figures.

Bidirectional, interactive TCP traffic in ns-2 requires the use of an alternative TCP implementation (called "FullTCP"). The ns-2 implementations of ADTCP and TCP-AP are based on the standard (unidirectional) TCP implementation, and are therefore not able to support HTTP traffic. We can thus give HTTP results only for TCP Newreno and the two BarRel variants.

In Figure 3.9, we increase the number of servers from one up to five. Since each server has three clients, this also increases the network load. The maximum node speed in all these simulations is 6 m/s. As expected, for both TCP and BarRel the mean page access time increases with an increasing server count. Both variants of BarRel exhibit a substantially lower page access time. Moreover, the degradation with increasing network load is more severe for TCP than for BarRel, particularly for the TRFA variant.

The traffic situation with many small requests and small data object transmissions is the worst case for the CaRe variant of BarRel. For a high load—particularly in the case of five servers—the overhead of the CaRe packets becomes visible. Nevertheless, despite the effort for the CaRe packets the performance is still substantially better than that of TCP.

In Figure 3.10, we show similar evaluation results, but now for a number of servers fixed at three and an increasing maximum node speed. Albeit a certain degradation with increasing node speed is visible, these results underline our finding from the FTP

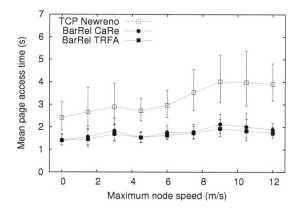

Figure 3.10: Mean HTTP request latency with varying node mobility.

simulations that node mobility has only limited impact on the performance of either protocol. Network load is the more critical factor.

Figure 3.11 is, like above Figure 3.3, a cumulative distribution function, here of the page access times for a maximum node speed of 6 m/s and three servers. The significant difference particularly at the tail of the distribution is evident: while virtually all requests are completed within ten seconds for the BarRel variants, more than 15 % of the TCP requests exceed this threshold.

3.4 Chapter Summary

In this chapter, we have introduced a transport protocol, BarRel, that builds upon CXCC. BarRel's fundamental design with respect to reliability provisioning—founded on implicit feedback—is vastly different from existing solutions. In particular, it is able to provide TCP-equivalent service, while avoiding oncoming control traffic. It therefore demonstrates new, previously unexplored options in the design space of protocols for wireless multihop networks.

We have introduced two mechanisms for BarRel to acknowledge the successful reception of the end of a packet burst. The first mechanism uses one single end-to-end ACK at the end of a packet burst, and TRFA packets to recover from packet losses at the end of a burst. The second mechanism uses CaRe packets to achieve reliable wireless multihop communication without any end-to-end acknowledgment traffic from the sink back to

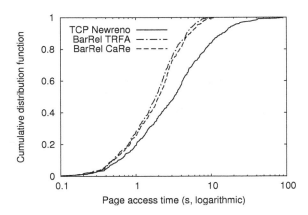

Figure 3.11: Cumulative distribution function of HTTP page access times.

the source. Since the number of CaRe packets that need to be added at the end of each burst increases with the route length, this latter mechanism implies a somewhat higher overhead. It has, however, other benefits; in particular, it is very straightforward, and unlike the TRFA-based mechanism it does not require choosing timeouts. In our simulation study its performance has been shown to be comparable to that of the TACK/TRFA-based approach.

Our simulations also generally underlined the superiority of the BarRel/CXCC protocol stack over TCP/802.11-based solutions in a wireless multihop environment. We have analyzed the performance for both unidirectional FTP-like data transfers and interactive, HTTP-like traffic, and have demonstrated the benefits of our approach with regard to many different performance metrics. These results support our claim that for the design of future wireless multihop networks alternatives to the common Internet protocol stack should be considered, particularly in those cases where application-specific networks are designed.

Chapter 4

Implicit Multicast Congestion Control: BMCC

Because of the shared medium and the resulting tight capacity limits for each node, multicast communication is of particular interest in wireless multihop networks: it helps saving resources when delivering data to multiple destinations. This is further strengthened by the fact that group communication is an inherent feature of many proposed applications for wireless multihop networks. But a shared broadcast medium is, as seen before, also much more prone to network congestion than, for example, traditional wireline networks. Reducing the number of transmissions required to deliver the data to all receivers is therefore only half the battle. It is, especially in wireless multihop networks, also absolutely vital to perform efficient congestion control, to avoid a congestion collapse.

We therefore now extend the concept of implicit hop-by-hop congestion control to the case of multicast communication. We propose a novel congestion control scheme for multicast in mobile ad-hoc networks. While generally exhibiting very competitive performance, it focuses particularly on the most demanding class of applications: those which depend on very low packet latencies in combination with high packet delivery ratios. We call this scheme *Backpressure Multicast Congestion Control* (BMCC). BMCC and the central results of this chapter are also covered in [STL+07].

We focus on an implementation of BMCC in combination with Scalable Position-Based Multicast (SPBM) [TFW+07], a geographic multicast routing protocol. Implementing congestion control over such a scheme is particularly challenging: the source node has neither information on the continuously changing topology of the multicast tree nor on the number of group members. BMCC can be used whenever forwarders know their set of next hop nodes in the multicast distribution tree. This holds for SPBM, but also for a large variety of other multicast routing approaches. Therefore, the ideas and concepts introduced here are not specific to SPBM.

In our evaluation, we assess the performance obtained with BMCC using ns-2 simulations. We compare it to plain SPBM, to a modified version of SPBM, and to the On-Demand Multicast Routing Protocol (ODMRP) [LGC99], a well-known topology-based multicast protocol for MANETs. The results of the simulations underline the very good performance of our approach.

4.1 Related Work

While multicast routing for mobile ad-hoc networks has received some attention over the last years, congestion control for this type of traffic in a wireless multihop environment has only been studied sporadically.

In [TG03] the MANET multicast protocol ODMRP is evaluated with a different MAC protocol than IEEE 802.11. In this approach congestion control is performed in an end-to-end fashion. Explicit notifications inform the sender about the average load on the used links. The authors argue that a backpressure mechanism would react too slowly. Our protocol, however, proves the opposite, reacting virtually immediately if the forwarding of a packet is delayed.

Similar to the above approach, Tang et al. [TOLG02a, TOLG02b, TOLG03] introduce an end-to-end congestion control protocol for multicast traffic. The authors propose to use negative acknowledgments to infer congestion. The sender reacts by reducing its rate until one affected receiver acknowledges a reception explicitly. Rajendran et al. [ROY$^+$04] also use end-to-end rate adaption. In addition, they anticipate upcoming congestion by a local repair strategy, reducing the amount of explicit congestion notifications. Both approaches, however, depend on feedback from the group members, causing a substantial amount of feedback traffic. Our protocol builds up backpressure immediately and locally, and avoids explicit feedback.

In [BZK04, Bau05], Baumung et al. propose a congestion controlled multicast overlay for MANETs. Hierarchical aggregation of acknowledgments provides feedback on the progress of the worst receiver to the source. This feedback is then leveraged for congestion control. This approach is well-suited for overlay multicast abstracting from the underlying network. However, detecting packet loss at the receivers and propagating the aggregated feedback may require significant time. This is avoided by our approach.

Peng and Sikdar propose a congestion control scheme for layered multicast in MANETs [PS03]. In their protocol, multicast layers are blocked and released in intermediate nodes, based on the observation of per-link output queue lengths and throughput measurements. Adjustments finer than a whole layer are thus not possible. The scheme also does not take all aspects of the shared medium into account: links are considered heterogeneous and lossy, but independent in terms of capacity. The packets of a blocked layer are still delivered to the blocking intermediate node, and dropped there. This wastes valuable shared medium capacity in the bottleneck area. These problems do not exist in our protocol.

There is substantial existing work in the area of MAC layer multicast in wireless environments. These do not deal with congestion control for multihop multicast traffic, but with delivering a packet to multiple (local) receivers. As one amongst other aspects, our scheme also has to deal with this question. A typical example is the Multicast MAC (MMAC) protocol [GNAA04]. In MMAC, the receivers of a transmission are listed in the packet header. Each of them explicitly acknowledges the successful reception, in the order given by their index position in the header. In [JD06], a scheme is introduced which transmits a data packet to up to four receivers at once, and collects acknowledgments from them; for more than four addressees, clusters of at most four nodes are formed, and the packet is transmitted to each cluster separately. This paper also provides a broader overview of the area. We consider single-hop delivery to multiple addressees in a larger context, conjointly with multihop backpressure. This allows for a different view of the problem. All previously proposed approaches result in significant control overhead like, e. g., round-robin polling of all destinations, or many additional feedback fields. This is not necessary in our approach.

4.2 Scalable Position-Based Multicast

Scalable Position-Based Multicast (SPBM) by Transier et al. [TFW+07] is a position-based multicast routing protocol for mobile ad-hoc networks. It consists of two main components: a group management protocol and a multicast forwarding protocol. Both use a subdivision of the network area according to a hierarchical quadtree, as shown in Figure 4.1. The whole area is divided into four sub-squares, which are in turn again subdivided, and so on. This is continued until the resulting so-called level-0 squares are small enough so that each node is able to communicate with all the other nodes within the same level-0 square directly, i. e., they are in a one-hop distance.

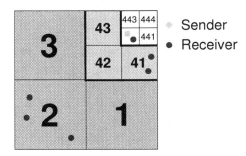

Figure 4.1: An example for multicast forwarding in SPBM.

The group management provides every node in the ad-hoc network with aggregated group membership information. For each of the three neighboring squares on each hierarchy level, a node knows the list of groups of which at least one member resides in the group. For example, a node located in square '442' in Figure 4.1 has knowledge about the aggregated membership information of squares '441', '443', '444', '41', '42', '43', '1', '2', and '3'. How this information is actually provided is described in detail in [TFW+07].

The forwarding decision is based on information about neighboring nodes. Each node maintains a table of nodes in its transmission range. This is accomplished by overhearing data messages and periodic update messages issued by the membership management service, which contain the ID and position of the sending node.

A packet that is to be forwarded includes a list of destination squares, and a group address indicating the group to which the packet is being sent. Again in the example in Figure 4.1, the source node in square '442' would address a packet to the one shown member in '442' directly (because it is located in the same level-0 square and is thus explicitly known), and to the squares '41' and '2' (because the membership management knows that group members reside there). Upon reception of a forwarded packet, a node checks whether it has more detailed information on the destinations. This will happen as soon as a packet enters its destination square. The respective entry is then *disaggregated* by the forwarder. For example, when the packet with destination '41' in Figure 4.1 enters square '41', the destination will be replaced by entries for squares '411' and '414': the forwarder, itself located within '41', knows that these are the sub-squares where the group members reside.

The packet is then handed over to the forwarding algorithm, where the best-suited neighbor to forward the packet to each of these destinations is identified. This is accomplished similar to position-based unicast routing (see [MWH01]): the source compares the geographic progress for each of the neighbors with respect to the destination and picks the neighbor with the greatest progress. After identifying the next hop for each destination, the forwarding algorithm sends a copy of the packet to each of these next hops, addressed to the respective destination field(s) that shall be reached. The forwarding uses a sequence of unicast transmissions. This increases the reliability, since there will be MAC layer acknowledgments. It comes, however, at the cost of multiple messages. In the following, we describe the adaptations we made to this transmission scheme in order to implement our congestion control algorithm.

4.3 Backpressure Multicast Congestion Control

The central difference between unicast and multicast from the perspective of packet forwarding is that for unicast each forwarder has exactly one next hop node, while with multicast there may be more than one. Essentially, each packet is forwarded along a tree of nodes, originated at the source node. Therefore, in order to apply implicit hop-by-hop congestion control to multicast traffic, we need to generalize the implicit feedback concepts appropriately to that situation. This generalization forms the core of BMCC.

4.3.1 Packet Forwarding with Local Broadcasts

Implicit hop-by-hop congestion control embraces the local broadcast property of the wireless multihop medium by not using explicit feedback, but instead gathering information through overhearing. Transmitting the payload of a packet directed to multiple next hops only once is thus a natural approach, and it is followed here just like in many previous proposals.

As a first step, we use a modified version of SPBM, called *Broadcast SPBM* (SPBM-BC). In SPBM-BC, the group management and the selection of the next hops are done as described in Section 4.2. But instead of a separate unicast transmission for each next hop, a single broadcast transmission is used for all of them. The forwarding node adds all designated next hops to the packet header, including a list of destination squares for each of them. When the packet is sent via MAC layer broadcast, all neighbors receive it and check whether they are contained in the list of designated next hops—if not, they will discard the packet. This is complemented by implicit acknowledgments: if the original sender does not overhear the retransmission of a packet from all of the designated next hops, it will rebroadcast the packet after removing all next hops that already successfully acknowledged the packet.

SPBM-BC is the basis for SPBM with Backpressure Multicast Congestion Control, but it will also serve as a benchmark: SPBM-BC will show us the performance that can be obtained by using local broadcasts and implicit acknowledgments *without* BMCC's backpressure mechanism.

To avoid parallel medium access attempts by multiple addressees, each attempting to forward the packet, a node waits for a random backoff before it transmits. Multiple next hop nodes may not all be within mutual communication range, but it is reasonable to assume that they are often within carrier sense range. In combination with carrier sensing and medium access backoff the jittering desynchronizes the answers, thus avoiding the synchronization problem. This pragmatic solution matches the soft timing principle of CXCC very well, avoids complex co-ordination, and saves significant overhead.

4.3.2 Backpressure with Multiple Next Hops

By not allowing the transmission of a subsequent packet before its predecessor has been forwarded by the next hop CXCC builds up backpressure. This guarantees that a downstream bottleneck rapidly propagates backwards along the route towards the source. In BMCC, we apply the same concept, but along a tree structure. We strive for high packet delivery ratios to all receivers in this tree, i.e., towards all leaves. As a consequence, we need to adjust the source data rate to the tightest bottleneck in the forwarding tree. In other words, we need to ensure that the data inflow into any branch does not exceed the bottleneck capacity within that branch.

In this form, the scheme will be susceptible to the well-known "crying baby problem" [HSC95]. If one group member has a particularly bad connection, its mere existence will result in a deterioration of service quality for the other group members. We will devise a way to deal with this effect later. For now we concentrate on a backpressure protocol that adjusts the source data rate to the tightest point in the multicast tree.

BMCC achieves the desired congestion controlling behavior by generalizing the CXCC backpressure rule in the following way: the next packet may only be forwarded if *all* the next hop nodes for that packet have forwarded the previous one. Similar to the backpressure building up backwards along the route with CXCC, this rule in BMCC results in backpressure along the tree. Thereby, packets that are not able to traverse the network will not be allowed to leave the source node. This implicitly regulates the source data rate, and it keeps the queues in the intermediate nodes extremely short. Each forwarder can queue at most one untransmitted packet. The source node can also communicate the backpressure to the application. This allows to adapt the packet generation to the medium situation, for example by adjusting the bit rate dynamically.

Since transmissions in BMCC are directed to a set of next hop nodes, the situation is significantly more complex than in the single next hop case of unicast forwarding. Each single next hop node may have received the transmission correctly or not. If the packet has been received correctly, each of the next hops may already have forwarded it again or might still hold it back due to backpressure. Finally, for each successor having forwarded the packet, the implicit acknowledgment may have been overheard or not. The central challenge in BMCC is to deal with this additional complexity efficiently while adhering to the principles of implicit feedback and soft timing, and avoiding unnecessary control traffic.

To tackle this challenge, a forwarding node in BMCC keeps track of the list of next hop nodes from which an acknowledgment is still missing. After transmitting a packet addressed to a set of one or more next hops, this list is initialized to contain all these next hops. If an implicit (or explicit) acknowledgment from one of them is detected, the respective node is removed from the list. The transmission of the subsequent packet is allowed once the last entry has been removed from the list.

If acknowledgments are missing for a too long time, a generalization of CXCC's RFAs is used. Just like data packets, RFAs in BMCC are directed to a whole set of next hop nodes: they address all the next hop nodes from which an acknowledgment is still missing. All thereby challenged forwarders can decide individually whether they should react with an explicit ACK or NACK.

A number of optimizations is possible to exploit the information contained in these handshakes most effectively. Since a single next hop node that has not received the data packet already necessitates a retransmission, it is not necessary to wait for feedback from all nodes if a NACK is received. In this case, an immediate retransmission of the data packet is triggered, addressed to the nodes from which acknowledgments are missing. Ideally, this makes the transmission of further NACKs by other next hop nodes unnecessary. Furthermore, such a retransmission may also fulfill the purpose of an RFA for nodes that had already received and forwarded the packet. If their forwarding has not been overheard by their predecessor, they will be in the list of addressees of the retransmission. They can easily detect this situation and repeat the lost feedback through an explicit ACK.

Like for the packet transmissions themselves, a possible synchronization of the answers to an RFA needs to be considered. If multiple addressees all access the medium immediately after receiving the RFA this will cause severe collisions. For this reason, such reactions by forwarders are, just like forwarded data packets, sent with substantial jitter.

4.3.3 Dealing with Unavailable Next Hops

In order to perform effective congestion control, backpressure should be maintained as long as the downstream nodes are not able to forward the previous packet. It must, however, be avoided to wait indefinitely for an implicit acknowledgment from a downstream node which is no longer reachable. Such a node will obviously not react to RFAs. But since this also applies to a node keeping a packet back due to backpressure, a lightweight mechanism is needed which helps to distinguish these two cases.

A basic solution to this problem is already provided by SPBM: if no more update beacons from a neighbor are received over some time, it is considered unavailable. But due to the relatively low beaconing frequency, this reacts rather slowly. We thus also adopt the KAL packet mechanism of CXCC for the detection of broken links, thereby speeding up the detection of no longer available next hop nodes.

Recall that a KAL packet is sent if an RFA is received for a packet which has arrived, but is currently held back due to backpressure. It may also be issued when a new packet is received from the previous hop, while an acknowledgment for the preceding one has not yet been received. It indicates that its sender is reachable, but it does not release the backpressure. With this extension, the link to a next hop node may be considered broken if the number of consecutive unanswered RFAs exceeds some threshold.

4.3.4 Handling Inhomogeneous Receivers: Backpressure Pruning

One earlier mentioned issue still deserves attention: BMCC will adjust the data rate to the tightest bottleneck in the multicast tree, i.e., according to the slowest receiver. While this is necessary in order to achieve high delivery ratios at all receivers—and might thus well be desirable in certain usage scenarios—, it is susceptible to the crying

Figure 4.2: Simple scenario with unequal receivers.

baby problem. If there is one group member with a particularly bad reachability, this will thwart a higher data rate to all other receivers. It is thus of interest to see whether a variant of BMCC can be built that exhibits a different behavior in this regard: is it possible to modify the algorithm to adjust the inflow into each branch of the multicast tree to the highest rate sustainable by at least one receiver in that branch, thus maximizing the throughput to each individual receiver? Depending on the application, the original version or such a variant may be favorable.

However, due to the shared broadcast medium the rates to the receivers cannot be individually and independently maximized. For clarification, let us consider two simple examples in a scenario like in Figure 4.2. There is one sender and two receivers. While receiver 1 is directly reachable from the source, receiver 2 is further away. In the first example, there is no additional traffic in the network. Transmissions from the source to receiver 1 are affected by transmissions made by at least the first two forwarders towards receiver 2, because of the shared medium and carrier sensing at the source node. BMCC aims at high packet delivery ratios to all receivers. The backpressure rule as presented above will achieve the following: it will always allow to forward a packet towards receiver 2 before the next packet enters the network—even though this reduces the throughput to receiver 1. When striving for fairness between multiple receivers and high packet delivery ratios this is generally the desired behavior. Ideally, receiver 1 should not receive data at a higher rate, if this comes at the cost of receiver 2's rate.[1]

As a second example, let us consider a situation in which the medium around receiver 2 is severely congested. Then, backpressure towards the source will build up, and forwarding of packets along the route to receiver 2 may be substantially delayed—this is inevitable if it is not possible to forward packets towards receiver 2 at a higher rate.

---

[1]This is actually related to the notion of max-min-fairness, which states that a resource allocation is max-min-fair if increasing the share of any component is only possible at the cost of decreasing the share of an already lower component (for an in-depth discussion in the networking context see, e. g., [RL02]). We do not claim that the variant of BMCC to be introduced now will guarantee max-min-fair bandwidth allocations—due to the complexity and stochastic nature of a wireless multihop environment such a guarantee is hardly possible. But we aim for a heuristic that follows this general idea.

But this behavior can result in substantial *underutilization* of the medium around the source and receiver 1. Depending on the application, it may be desirable to use such otherwise unused medium bandwidth for the forwarding of additional packets to better reachable receivers. Nonetheless only those packets should enter the network for which the bandwidth towards at least one receiver suffices.

We will now present a modification of the backpressure rule of BMCC that is able to yield just these effects. We call it BMCC with *backpressure pruning* (BMCC-BP). The backpressure pruning mechanism allows for branches to be cut off if backpressure exists in them. It makes use of the keepalive packets introduced above for the purpose of improved unavailable forwarder detection. A KAL packet occurs in backpressure situations, when the forwarding of packets is delayed. Thus, the reception of a KAL from one next hop node indicates that the respective subtree is currently a bottleneck.

In standard BMCC, a node must wait for all next hop nodes to acknowledge the packet (neglecting, for simplicity of discussion, possible unavailable next hop nodes). BMCC-BP replaces this with a slightly more complex rule set, as follows. A node may stop further attempts to deliver a packet to the remaining next hop nodes if

1. at least one next hop node has acknowledged the packet,

2. a KAL has been received from all other next hops, and

3. a subsequent packet is already available for forwarding.

At the source node, the latter criterion is fulfilled if the application has already generated a subsequent packet which is waiting in the queue. In intermediate nodes, it holds as soon as a follow-up packet has been received from the upstream node. This may happen when the upstream node, in turn, has received at least one implicit or explicit ACK and KALs from all *its* remaining next hops.

The first backpressure pruning criterion guarantees that each packet will eventually arrive at at least one receiver: if one next hop node has acknowledged the packet, this implies that it has been forwarded into at least one branch. Packets will thus still not enter the network at a rate higher than what can be sustained by the "best" group members. The second criterion antagonizes the "stealing" of bandwidth from other branches, by providing each next hop with a chance to access the medium and thus at least with an opportunity to forward a packet.

Backpressure pruning may result in situations where a node receives a follow-up packet before it has attempted to forward the previous one. In this case, it should drop the previously known packet (for which it had sent a KAL), and instead enqueue the newly received one.

Summarizing so far, BMCC, as originally introduced, is designed to result in an adjustment of the source data rate to the tightest bottleneck in the network. This ensures high delivery ratios whenever possible. BMCC-BP is in some sense complementary: it is built to deliver the maximum individually sustainable rate to each receiver, as long as this does not come at the cost of other branches of the tree. Both adjust the rates of the source and of intermediate nodes without explicit rate feedback and without multihop control packets, by using implicit backpressure.

4.4 Evaluation

For the evaluation we implemented SPBM with BMCC in ns-2 [ns2a], based on Transier et al.'s SPBM implementation from [TFW+07]. As a comparison, we used the plain (unicast) version of SPBM as outlined in Section 4.2, the broadcast version described in Section 4.3.1, and an implementation of ODMRP [LGC99] that was originally obtained from [ns2b], ported to ns-2.30, and optimized as described in [TFW+07]. To obtain the results presented below, we used parameters similar to the ones in the previous chapter, with a network area of size 1500×1500 meters and a total of 200 nodes.

One multicast group was defined, with two senders and ten receivers; thus, two independent multicast trees were used in parallel. Other combinations of sender and receiver counts yield similar results. We consider scenarios with and without node mobility. The source applications generate data packets with 64 bytes of payload at an increasing rate between 1 and 50 packets per second, or the highest frequency at which packets are able to leave the source node, whichever is lower. While BMCC can provide fine-grained feedback to the application about when packets may be sent, the other protocols used here are not able to generate such feedback. By accounting only for packets that are able to leave the source, we thus avoid distortions of the results for the other protocols and keep the comparison fair.

Again the error bars in our figures show 95 % confidence intervals, based on (log)normal and i. i. d. sample assumptions. While, for the reasons discussed before, we use geometric means in our packet latency figures, we resort to the arithmetic mean for packet delivery ratio and throughput. The reason is that for both samples with value zero are possible (and do, for some protocols, actually occur). A throughput of zero was not possible

in our BarRel simulations described in Section 3.3, where all connections had a given
amount of data to transmit with no time limit imposed.

4.4.1 Delivery Ratio and Throughput

In Figure 4.3, the packet delivery ratio achieved by the different protocols is shown, in a
static setting without mobility. A value of one means that all packets that had left the
source nodes arrived successfully at each receiver. Adjusting the source rate in order to
allow for high packet delivery ratios was a main design goal of BMCC. The results show
that it has been achieved. At packet generation rates below 20 packets per second the
delivery ratios of unicast SPBM and ODMRP are very similar, with slight advantages
for SPBM. For higher rates, ODMRP delivers a greater fraction of the packets. However,
the fact that the delivery ratio is continuously decreasing for all three shows that more
and more packets are able to leave the sources, but then do not make it to the receivers.
Interestingly, the broadcast version of SPBM with implicit acknowledgments does not
reach the performance of unicast SPBM—although it theoretically needs fewer packet
transmissions. If the forwarding of a packet in one of the next hop nodes is delayed, no
explicit acknowledgment will arrive, which causes SPBM-BC to unnecessarily retransmit
the packet. This wastes substantial medium bandwidth. Obviously, it is not enough to
use implicit acknowledgments at the network layer. It is the backpressure mechanism
in BMCC that turns the balance. It is able to outperform all others and reach delivery
ratios very close to 100 % at all sending rates. This shows that the congestion control
mechanism is successful in regulating the source rate. The sending nodes only put on
air as many packets as the network is able to handle. Thus, the protocol is able to retain
high delivery ratios even for high packet generation rates.

In Figure 4.3, there is only a very limited difference between the results with standard
BMCC and BMCC-BP. This will likewise be the case in all our other random topology
simulations. As we will soon demonstrate, the reason is not that the protocols do
generally behave identically. This effect is rather caused by the relative homogeneity of
the receivers and therefore the load distribution in these networks. There seems to be
virtually no free medium capacity that could be used to deliver data faster to certain
receivers, without at the same time negatively affecting others.

Figure 4.4 shows how the packet delivery ratio develops in the presence of mobility.
In the mobile scenario simulations, the nodes move according to the random waypoint
mobility model, with a maximum speed of 5 meters per second, a minimum speed

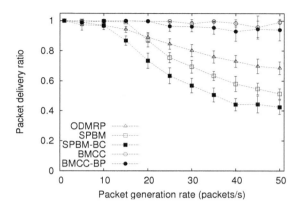

Figure 4.3: Packet delivery ratio with increasing packet generation rate.

of 0.5 meters per second, and no pause times. Initial node positions and speeds are sampled according to the mobility model's stationary distribution [NC04]. Data is again generated by the source applications at varying rates. It can be seen that the relative performance of the protocols remains largely unchanged, all deal reasonably well with mobility. Rapid topology changes cause inconsistencies in SPBM's routing tables, and thus also affect SPBM with BMCC.

A high packet delivery ratio could of course be achieved relatively easily if the total number of packets in the network was kept at a low level. Figure 4.3 does only show that almost all out of a so far unknown number of packets leaving the source do arrive with BMCC. We therefore have to consider these results in conjunction with the obtained data rate. Figure 4.5 presents the data rate achieved by an average sender-receiver pair. For packet generation rates of up to 10 to 15 packets per second, all examined protocols are able to deliver all the data produced by the applications. Since each data packet carries 64 bytes of payload, the resulting optimal data rate is 640 Byte/s at 10 packets per second. The simple broadcast version of SPBM breaks first. Starting from 10 packets per second, its goodput increases much less than the data generation rate. Plain SPBM and BMCC show similar trends at different levels: the goodput grows up to a certain saturation and stays at this level for higher packet generation rates. ODMRP delivers higher data rates starting from 30 packets per second. This, however, comes at a high cost: ODMRP then loses, as seen before, at least 20 % of the packets. As we will soon demonstrate, it also allocates resources unfairly, preferring close-by receivers, burdens the network with a heavy traffic load, and suffers from high delays.

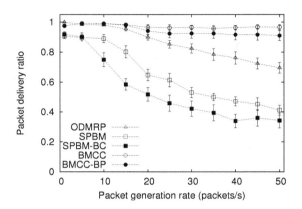

Figure 4.4: Packet delivery ratio in mobile scenarios.

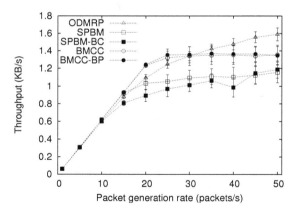

Figure 4.5: Throughput per sender-receiver pair.

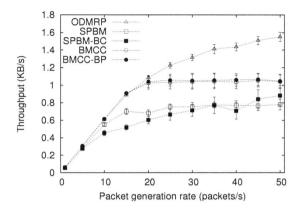

Figure 4.6: Throughput per sender-receiver pair in mobile scenarios.

Figure 4.6 shows the average receiver data for mobile scenarios. Again, BMCC achieves a perfectly shaped throughput curve. As can be seen, ODMRP is barely affected by mobility. The variants of SPBM, including BMCC, achieve lower rates in the presence of mobility. Starting from the point where ODMRP achieves higher data rates, it also—as described above—exhibits decreasing packet delivery ratios. In [TFW$^+$07], it has already been shown that SPBM suffers from mobility, because of its group management. Nevertheless, with BMCC it is able to keep up high delivery ratios.

4.4.2 Fairness Between Senders

The previous evaluation raises the question why BMCC does not achieve the somewhat higher data rates obtained with ODMRP, if the network is seemingly able to support them. The key to understand this property lies in the vastly different effort that is required to deliver a packet to different receivers, depending on their distance from the source. It is much more resource intensive to bring a packet to a far away receiver than to a close-by one. A high data rate might simply be obtained by preferring transmissions over shorter distances. This issue is closely related to the fairness between senders: do receivers preferably receive packets from closer source nodes?

In order to analyze this aspect, we look at the distribution of packet sources amongst the packets arriving at the receivers, with an increasing number of senders in a multicast group. To quantify the fairness of this distribution, we use Jain's fairness index as introduced in [JCH84]. This index establishes a measure for the fairness of resource

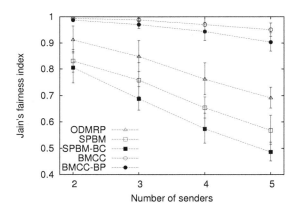

Figure 4.7: Jain's fairness index for packet distribution over senders.

allocation in a multi-user system. It yields a value between zero and one, where one means perfect fairness and zero is approached if one out of more and more participants is assigned all resources. The fairness index is defined as

$$\frac{\left(\sum_{i=1}^{n} x_i\right)^2}{n \cdot \sum_{i=1}^{n} x_i^2},$$

where x_i is the resource share assigned to the i-th participant.

Here, we apply Jain's fairness index to the packet counts received from each source, giving us a fairness value for each receiver. I. e., in our case x_i is the number of packets received from the i-th source. In Figure 4.7, the average of the resulting index values for all receivers is shown, for an increasing number of senders. ODMRP achieves a better fairness than plain unicast SPBM. Broadcast SPBM does not meet the performance of the unicast version in this metric either. But BMCC again clearly outperforms ODMRP. Here lies the reason why BMCC does not allow for higher data rates: these seem possible only at the cost of an increased unfairness.

4.4.3 Delay and Protocol Overhead

Figure 4.8 shows the average total amount of data that has been transmitted on the physical layer during one simulation run, as a measure of the protocol overhead. ODMRP has the highest resource requirements. The reasons lie within the structure of the protocol: ODMRP floods data packets through the whole network on a regular basis, and

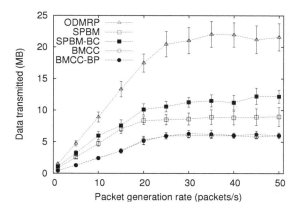

Figure 4.8: Data transmitted on physical layer.

it uses redundant paths in a mesh structure, both of which result in a higher number of transmissions. The broadcast and unicast version of SPBM produce similar amounts of data on the physical layer. If local broadcasts are used with SPBM without employing BMCC's backpressure mechanism, it needs even somewhat more bandwidth, instead of saving it. This results from a high number of retransmissions performed by this approach. The backpressure mechanism of BMCC, because of its effective ways to avoid unnecessary retransmissions, is once again able to turn this into the opposite, avoiding unnecessary control traffic and retransmissions.

Finally, we look at the end-to-end delay. Figure 4.9 depicts the geometric mean of the end-to-end delay of all delivered data packets, from the time the packet leaves the source node until it arrives at the receiver. Again, up to a packet generation rate of 15 packets per second in each source, all protocols deliver the packets at sufficiently low delays. For higher data generation rates, only BMCC is able to maintain short packet latencies. The other protocols delay packets for up to several seconds, which is definitely unacceptable. This problem stems from long queues building up in the intermediate nodes, a problem being avoided in BMCC by the very design of the protocol, which implies very short queues.

4.4.4 Backpressure Pruning

So far, it seemed that backpressure pruning does not have any substantial effect. This is, however, not true. The impression is a result of the relative homogeneity of the so

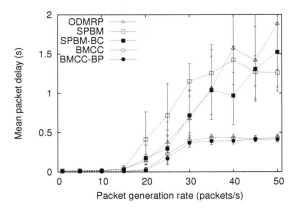

Figure 4.9: End-to-end delay.

far considered settings. To analyze the behavior of BMCC-BP in a scenario in which a difference must clearly appear, we use a simple static topology similar to the one depicted in Figure 4.2. Based on an equidistant chain topology, the nodes are set up such that the source is a direct neighbor to one receiver, R_1, while a second one, R_2 is seven hops away. An additional interfering data stream transmits packets continuously in the neighborhood of this second receiver. The source node again generates data packets at an increasing rate.

We analyze the packet delivery ratio as well as the data rate for both receivers separately. This allows for a detailed analysis of the operation of BMCC-BP. The respective results are depicted in Figures 4.10 and 4.11. There are six curves, describing the results with ODMRP and the two BMCC variants. For improved readability of the charts we leave out the results with SPBM. Not surprisingly, all protocols are able to transmit packets with a high delivery ratio to the first receiver. BMCC, aiming at the maximum possible fairness, notices the congested area via its implicit backpressure mechanism and thus maintains a high delivery ratio also towards the second receiver—which is only possible at a limited rate for *both* receivers. ODMRP, without a mechanism to deal with such a congestion situation, results in a high number of lost packets on the path to the second receiver; the first destination receives packets at a high data rate, while the second receiver is mostly cut off. The backpressure pruning mechanism in BMCC-BP handles the congestion situation correctly. It reduces the data rate to the second receiver (and thus the packet delivery ratio) without affecting the ability of the non-congested receiver to receive more packets.

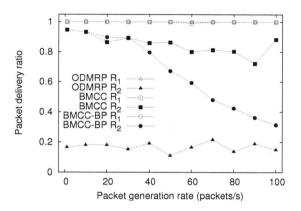

Figure 4.10: Packet delivery ratio for receivers R_1 and R_2 in a simple static scenario with congestion at R_2.

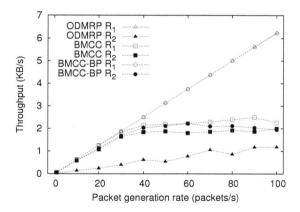

Figure 4.11: Data rates of receivers R_1 and R_2 in a simple static scenario with congestion at R_2.

4.5 Chapter Summary

In this chapter, we have proposed a novel way to perform effective congestion control for multicast traffic in wireless multihop networks. Our scheme is based on implicit feedback, establishing multihop backpressure through simple packet forwarding rules. It solves single hop reliability and implicit multihop backpressure congestion control conjointly, thereby avoiding many unnecessary control messages and packet retransmissions.

A simulator implementation of the approach in combination with the geographic multicast routing protocol SPBM exhibits convincing performance in a simulation study, demonstrating the effectiveness of the source rate limitation. Our scheme yields competitive throughput and superior fairness while maintaining very high packet delivery ratios for all receivers, combining these traits with very low end-to-end packet delays and a low protocol overhead.

Chapter 5

Co-ordinated Network Coding: noCoCo

With network coding, a router transmits multiple packets within a single "coded" packet, and can thereby make more efficient use of the available bandwidth. The recent work COPE by Katti et al. [KRH+06] applies this technique to wireless multihop networks. Considering a simple three-node, two-hop setting outlined in [WCK05], the basic principle of network coding is easy to understand: if node A sends a packet p_A to C via B, and C sends a packet p_C to A, also via B, then B may send the XOR of p_A and p_C, $p_A \oplus p_C$, instead of transmitting them separately. Since A and C know p_A and p_C respectively, they can extract the data intended for them with another XOR operation: $p_A \oplus (p_A \oplus p_C)$ to retrieve p_C, and similarly $p_C \oplus (p_A \oplus p_C)$ to retrieve p_A. This is schematically shown in Figure 5.1: while without network coding, as in Figure 5.1(a), four transmissions are necessary, three suffice if network coding is used as in Figure 5.1(b). The concept can be transferred to more complex scenarios; COPE generalizes it to using overheard packets to decode.

In COPE, coding is performed when opportunities arise spontaneously. Such an opportunistic approach may work very well if, in the above example, indeed both A and C make their transmissions before B accesses the medium. Then, B has packets p_A directed to C and p_C directed to A in its queue. It also is aware that A knows p_A and C knows p_C, because these are the nodes from which the respective packet has been received. B can thus seize the opportunity to reduce the number of required transmissions by transmitting $p_A \oplus p_C$, knowing that both receivers will be able to extract their respective packets.

Not always, however, will things work out that well. If A transmits first and B forwards the packet further on immediately, no coding will be performed. Consequently, the order of transmissions can significantly impact the availability of coding opportunities, and hence the coding benefits.

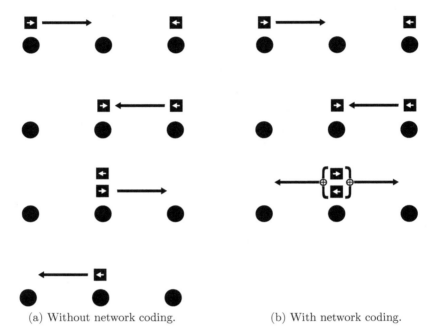

(a) Without network coding.   (b) With network coding.

Figure 5.1: Exchange of two packets over a wireless relay.

This prompts us to investigate more deterministic alternative approaches to coding. Network coding in the way performed by COPE builds upon the local broadcast property of the wireless multihop medium. It is therefore tailored to the same environment as CXCC, BarRel, or BMCC, and it likewise explicitly takes advantage of the network's characteristics. Hence, the question arises whether implicit feedback and co-ordination along the lines of the approaches pursued here can be applied to increase the number of coding opportunities, and thereby the efficiency of the medium use.

In the following, we start by analyzing packet flows in two-way data traffic, and conclude that it is possible to guarantee coding opportunities through carefully co-ordinating packets transmission. We introduce a centrally scheduled coding scheme and derive general properties of schedules that can maximize the coding gain in two-way traffic flows. After studying the theoretical limits, we translate the understanding into a practical protocol proposal, *Near-Optimal Co-ordinated Coding* (noCoCo). noCoCo is built around a few simple rules for packet forwarding that—even though they are designed for and motivated by a very different purpose—can be seen as an extension of CXCC. noCoCo guarantees achieving the maximum possible coding gain for each single bidirectional connection in isolation. It can be further combined with opportunistic schemes to identify and exploit additional coding opportunities with, e. g., cross-traffic or unidirectional traffic components. The results of this chapter are also the topic of [SHC07].

5.1 Related Work

Network coding was initially proposed by Ahlswede et al. in a theoretical work on multicast communications in wireline networks [ACLY00]. In this context, it was shown to substantially increase the network capacity. Network coding for unicast flows has subsequently been studied, also mostly from a theoretical and modelling perspective. Few of these works, though, consider a wireless context. Examples are [LL04, WCK05, HCH06]. Here, we implement network coding for two-way unicast flows in a distributed protocol for wireless multihop networks.

Wu et al. [WCK05] studied bidirectional traffic crossing a single relay. They introduced the concept to combine packets that traverse a relay in opposite directions by XOR. Katti et al. [KRH+06] generalized that and applied it in a practical, distributed protocol named COPE. It is implemented as a coding shim between the MAC and routing/IP layers. COPE may be considered the previous work most relevant to ours.

Beyond combining packets that traverse the same relay in opposite directions, COPE uses overheard packets for additional coding opportunities. For a coded transmission to be successful, all intended nexthop receivers must be able to decode it. This requires a receiver to have all other component packets except the one directed to itself. Consequently, the task of identifying coding opportunities boils down to obtaining information about which packets are known by which neighboring node. COPE employs three mechanisms. First, a node knows all packets it has sent out. Therefore the node from which a packet has been received will know that packet—this is the case discussed previously. Secondly, COPE nodes piggyback "reception reports" onto their transmissions, to explicitly notify neighbors of the packets received/overheard. Finally, COPE also "guesses" coding opportunities: based on the link quality information from the routing protocol, a node estimates the overhearing probabilities of coding candidates at specific neighbors. This yields the probabilities of successful decoding on all addressees for a particular combination of packets. If this success probability is above a threshold, the packets are combined into one packet—at the risk of occasional failed decoding on some neighbor.

Generally, the coding gain that can be achieved depends on the combinations of packets that meet in intermediate nodes, and thus on the routes chosen for the flows in the network. Some recent efforts considered cross-layer approaches in the context of coding-aware routing [SRB07]. Chaporkar and Proutiere also studied the issue of joint scheduling and COPE-like coding, focusing on characterizing the capacity region of a simplified coding scheme [CP07]. Here, we assume that oppositely directed traffic uses the same route. The substantial benefits that will be shown for our co-ordinated coding approach in fact motivate coding-aware routing that aims to achieve this property.

5.2 Maximizing the Coding Gain

5.2.1 A Centralized Scheduler

We concentrate on the non-opportunistic coding of packets belonging to the same bidirectional connection, consisting of the two flows from endpoint A to B, and from endpoint B to A. For our discussion, we consider a single bidirectional connection in isolation. We term the two endpoints A and B the "left hand side node" and the "right hand side node". In practice, and in our noCoCo implementation introduced later on, opportunistic coding is used to exploit additional coding opportunities between different connections.

Note that two-way traffic is the prototypical situation for possible continuous coding gain. Such traffic exists in many applications, including many forms of real-time communications. It has even been argued that symmetry of outgoing and incoming packets counts should be a design criterion for good protocols [KWC+05].

To obtain a maximum number of coding opportunities, we look at the scheduling of transmissions: in which order should which nodes transmit which data packets in which coded combinations? The first question that arises in this context is what the maximum coding gain is, after all. The follow-up question whether it can be achieved is intimately related: is there a schedule that can always guarantee maximum coding gain, over an arbitrarily long timespan? In the following, we will first show that optimal scheduling of two-way traffic is indeed possible, by explicitly giving a centrally scheduled solution. We then point out some interesting properties that *any* schedule with maximum coding gain will necessarily possess.

We first observe that the initial transmission of a packet, when it leaves the source node, can not be coded. This is because no other node in the network knows this packet, and would thus be able to undo the coding. In the ideal case, all other transmissions combine two oppositely directed packets. One such transmission will then yield two single-hop packet deliveries. At most two (oppositely directed) packets from the same connection can be combined. Therefore, if we manage to combine two packets in each transmission of an intermediate node, we obtain the maximum coding gain.

A close look soon reveals that for a connection over more than two hops, a coding partner for each transmission cannot possibly be available from the beginning. The intermediate nodes first need to have packets in both directions available. Let us thus consider the situation after some initialization has taken place. Enumerate all intermediate nodes along the route, starting at one. Assume that the initialization manages to place exactly one packet for either direction in each intermediate node with an odd index. If the total number of hops is odd, thus giving an even number of intermediate nodes, it places an additional left-directed packet in the rightmost intermediate node. Any other nodes with even indices hold no packets. Now further assume that we have a global, centralized scheduler available. We may thus arbitrarily decide on the occurring transmissions and their order. Let the scheduler work as follows:

1. First, all intermediate nodes with odd indices makes one coded transmission each, thereby forwarding one packet in each direction. These transmissions may happen in an arbitrary order.

2. If one or both of the end nodes have received a packet during Step 1, they "answer" by injecting a new packet into the network.

3. Then, all intermediate nodes with even indices will have a packet in each direction available, and may now make coded transmissions.

4. Again, the end nodes reply to received packets by injecting a new packet.

5. Repeat this sequence of transmissions from Step 1.

For the case of a four-node, three-hop scenario, this is schematically visualized in Figure 5.2.

This scheme can be run indefinitely. All transmissions by intermediate nodes will always forward one packet in either direction, thus realizing optimal coding gain. This demonstrates that optimal coding gain is possible, though for now we do not know whether and how it can be achieved in a distributed way.

5.2.2 Notation

Before further examining the general properties of scheduling schemes that can guarantee successful network coding, we introduce a notation for the state of the network. The state of the queue is denoted in a node as x/y, where x is the number of queued packets directed to the node's left neighbor, and y the number of packets directed to the right neighbor. x, y are non-negative integers. We neglect all packets that do not belong to the connection under consideration.

To denote sets of possible states, we will use three placeholders. "$*$" means that at the respective position there may be any arbitrary non-negative integer, i.e., $*/y = \{x/y \mid x \in \mathbb{N}\}$. "$+$" stands for a positive integer, so $+/y$ denotes the set of states $\{x/y \mid x \in \mathbb{N}^+\}$. Finally, "?" means either 0 or 1, so $?/y = \{x/y \mid x \in \{0,1\}\}$.

If we look at a set of consecutive nodes along the route, we may write their joint queue state as $x_1/y_1, \ x_2/y_2, \ x_3/y_3, \dots$. Transmissions can then be expressed as transformations of the joint queue state. For example, a right-directed transmission of a single packet is

$$x_k/y_k, \ x_{k+1}/y_{k+1} \quad \rightarrow \quad x_k/y_k - 1, \ x_{k+1}/y_{k+1} + 1. \tag{5.1}$$

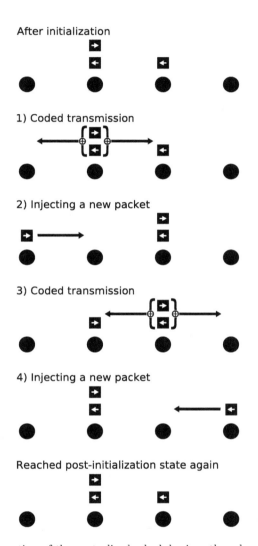

Figure 5.2: Operation of the centralized scheduler in a three-hop environment.

Such a transmission can take place if the transmitting node's queue state is in $*/+$.

Likewise, a coded transmission of two packets by node k would be

$$
\begin{aligned}
& x_{k-1}/y_{k-1},\ x_k/y_k,\ x_{k+1}/y_{k+1} \\
& \quad\to\quad x_{k-1}+1/y_{k-1},\ x_k-1/y_k-1,\ x_{k+1}/y_{k+1}+1.
\end{aligned}
\tag{5.2}
$$

The necessary precondition here is that the sending node's state x_k/y_k is in $+/+$.

5.2.3 Properties of High Coding Gain Schedules

Ideally, we want all transmissions of intermediate nodes to be coded, so we allow only transmissions as in (5.2) in the intermediate nodes. As mentioned earlier, no coded transmission can take place at the end nodes. Additional packets are thus inserted via uncoded transmissions. This yields state transitions of the following form at the leftmost intermediate node of the route:

$$
x/y \quad\to\quad x/y+1,
\tag{5.3}
$$

and equivalent ones at the rightmost intermediate node.

Packets leave the network by coded transmissions. For the leftmost pair of intermediate nodes, the corresponding transition is

$$
x_1/y_1,\ x_2/y_2 \quad\to\quad x_1-1/y_1-1,\ x_2/y_2+1,
\tag{5.4}
$$

with precondition $x_1/y_1 \in +/+$.

The availability of coding partners is likely if there are many packets in both directions available in the intermediate nodes. Conversely, coding opportunities will be rare if "too few" packets are on their way. However, as our results with CXCC, BarRel, and BMCC have demonstrated, it is desirable to keep the queues in a wireless multihop network short. Therefore, a suitable scheme needs to strike a balance between coding opportunities, throughput, and packet delay. We thus now determine how many packets are necessary in order to obtain high coding gains.

As a step in this direction, consider the allowable transitions (5.2), (5.3), and (5.4) and how they can be applied to "navigate" the global joint queue state space. We assume the use of some arbitrary scheduling scheme that maximizes the number of coding opportunities, and derive some properties that such a scheme must exhibit. It turns out that there is a subset of the state space which can never be left, should any of these states ever be entered. The following lemma points out this "state trap".

Lemma 5.1. *The subset of global joint queue states where the joint queue state of two consecutive nodes is in* $0/*$, $*/0$ *can never be left.*

Proof Consider the case where the joint queue state of a pair A, B of consecutive intermediate nodes is in $0/*$, $*/0$. Then, neither A nor B can make a coded transmission. A, however, can increase its number of left-directed packets only if a packet is received from B, and vice versa. Hence, this set of states can never be left by coded transmissions only. $\square$

Once such a state has been reached, it is impossible to maintain network coding with maximum gain—the affected nodes cannot continue to forward data with purely coded transmissions. This permits the reversal conclusion that a state as in Lemma 5.1 will never be reached by *any* scheme that maintains optimal coding gain.

If the joint queue state of each pair of neighbored intermediate nodes will never be in $0/*$, $*/0$, there always has to be at least one packet in one of the queues of each such node pair. The following theorem uses this to establish a lower bound on the number of packets in transit.

Theorem 5.2. *In any n consecutive intermediate nodes there are at least $n - 1$ queued packets at any point in time.*

Proof Let the joint queue state of the n nodes be $x_1/y_1, \ldots, x_n/y_n$. For each pair of consecutive nodes with states x_i/y_i, x_{i+1}/y_{i+1}, $1 \le i < n$, there is, as a consequence of Lemma 5.1, either $x_i > 0$ or $y_{i+1} > 0$. Therefore $\forall i, 1 \le i < n : x_i + y_{i+1} > 0$. For the total number of queued packets we obtain

$$
\begin{aligned}
\sum_{i=1}^{n}(x_i + y_i) &= y_1 + \sum_{i=1}^{n-1}(x_i + y_{i+1}) + x_n \\
&\ge \sum_{i=1}^{n-1}(x_i + y_{i+1}) \\
&\ge n - 1.
\end{aligned}
\tag{5.5}
$$

$\square$

This result can now immediately be used to obtain a lower limit on the total number of packets on the route, as follows.

Corollary 5.3. *For a bidirectional connection over h hops, there are at least $h - 2$ packets in the network at any point in time. This number must be exceeded temporarily.*

Proof The first assertion follows immediately from the previous theorem, since there are $h - 1$ intermediate nodes.

For the second part, consider (5.5) in the proof above in a situation just before a packet leaves the network. For this to happen, one of the outmost intermediate nodes must be in a state in $+/+$. Therefore, $y_1 > 0$ or $x_n > 0$. Then, the first inequality in (5.5) becomes strict, and thus the total number of packets is at least $h - 1$. $\qquad\square$

Let us see how close the centralized scheduler approaches this lower bound. For a route over h hops, it will, after the initialization, start with h packets in the network. Since a new packet is only injected after another one has left, this number is never exceeded. Therefore, the number of packets will always stay within the range $[h-2, h]$, and is thus nearly optimal.

It is worth mentioning that the bound in Corollary 5.3 is rather optimistic, and sometimes h packets are needed. Consider a two-hop connection over three nodes, i.e., the case $h = 2$. In order to perform a coded transmission, the middle node must have two packets available. Therefore, the number of packets in the network must be up to h in this situation.

In summary, it is indeed possible to schedule the transmissions in a bidirectional connection in a way that allows for the highest possible coding gain. There are global joint queue states which do not allow the intermediate nodes to proceed with only coded transmissions. Thus, an optimal schedule will have to avoid these states. This suggests that, when the number of packets being queued along the route is below a certain threshold, the coding cannot be optimal. Since a large number of queued packets in the network will increase packet delivery latency, there is a potential tradeoff between achieving the maximum possible coding gain and maintaining a low number of packets and thus short queues in the network.

5.3 A Practical Protocol

5.3.1 Basic Protocol Rules and Mechanisms

We will now refine the centralized approach from the previous section to obtain a practical and distributed scheduling scheme for network coding with success guarantees. We call this scheme *Near-Optimal Co-ordinated Coding (noCoCo)*. The key idea is to approximate the centrally enforced ordering of the transmissions in a decentralized way, based on implicit feeback. We generate a similar pattern of alternating transmissions of nodes with odd and even positions, while keeping the number of packets in the network as low as possible.

First, however, we need to initialize the network to a a valid starting state. Obviously, we need to begin with transmitting single packets as in (5.1). In our scheme, a node is allowed to forward single packets until it has "seen" packets going in both directions. Thereafter, only transmissions as in (5.2) and, at the outmost intermediate nodes, (5.3) and (5.4) are permitted.

The number of packets forwarded without coding should clearly be as low as possible. This motivates a backpressure rule similar to the one on which implicit hop-by-hop congestion control is based: a packet may only be transmitted to a node which has currently no packet for the same direction in its queue. This prevents an excessive number of packets from entering the network.

With the addition of the backpressure rule, the allowable state transitions can be refined in the following way. For uncoded transmissions by nodes that have not yet encountered packets in both directions, replacing (5.1), we get

$$x_k/1, \ x_{k+1}/0 \quad \rightarrow \quad x_k/0, \ x_{k+1}/1 \tag{5.6}$$

for a right-directed transmission by such a node. A coded transmission may only take place if the target queues in the neighboring nodes are free. Thus (5.2) becomes

$$0/y_{k-1}, \ 1/1, \ x_{k+1}/0 \rightarrow \quad 1/y_{k-1}, \ 0/0, \ x_{k+1}/1. \tag{5.7}$$

For packets entering and leaving the network, e. g., at the left end of the route, we get

$$x/0 \quad \rightarrow \quad x/1 \tag{5.8}$$

and

$$1/1, \ x_2/0 \quad \rightarrow \quad 0/0, \ x_2/1, \tag{5.9}$$

as refinements of (5.3) and (5.4), respectively.

By overhearing the forwarding of the downstream node, it can be verified that the packet has left the queue of the successor. Essentially, this results in CXCC-like forwarding during the initialization phase. Once packets in both directions have been heard by a node, the situation is even simpler: then, only coded packet transmissions are allowed. Thus, the previously transmitted packet has left the neighboring node if and only if the next packet from this node has been received.

In the beginning, packets from both sources will travel into the network. At most one packet from each source may be queued in each intermediate node. Eventually, there will be one node which is the first to hold packets in both directions. Let k be the index of this node. When such a node has emerged, the joint queue state of all intermediate nodes is in

$$0/?, \ldots, \ 0/?, \ \underbrace{1/1}_{\text{node } k}, \ ?/0, \ldots, \ ?/0. \tag{5.10}$$

The uncoded transmissions according to (5.6) allow each position with the placeholder ? in (5.10) to hold a packet. Taking this into account, as well as the fact that node k is allowed to perform a coded transmission, the state will eventually reach

$$0/?, \ldots, \ 0/?, \ 1/1, \ \underbrace{0/0}_{\text{node } k}, \ 1/1, \ ?/0, \ldots, \ ?/0. \tag{5.11}$$

Iterating this, it is easy to see that a situation will emerge, in which each node is allowed to transmit exactly once whenever both its neighbors have performed a transmission. This also holds for the end nodes. Hence, the emerging scheme is in fact very similar to the centrally scheduled scheme discussed before.

5.3.2 An Upper Bound on the Number of Packets

We will now look at the number of packets in transit for the proposed protocol. Note that this analysis shows a very interesting symmetry to the derivation of the respective general properties in Section 5.2.3.

Lemma 5.4. *With the proposed protocol, a state where the joint queue state of two consecutive nodes is in $+/*$, $*/+$ will never be reached.*

Proof For some node A with queue state x_A/y_A, due to the backpressure rule, it always holds that $x_A, y_A \leq 1$. Consequently, after A has performed a transmission, $x_A = y_A = 0$. The latter is true both during the initialization phase and during normal operation.

Assume the joint queue state x_A/y_A, x_B/y_B of a pair A, B of consecutive intermediate nodes is in $+/*$, $*/+$. Then, $x_A > 0$ and $y_B > 0$. $x_A > 0$ means that B's last transmission must have been more recent than A's last transmission. $y_B > 0$, however, requires the opposite, that A has transmitted more recently than B. This is a contradiction, hence the assertion holds. $\qquad\square$

This lemma can, very similar to what we did above in Section 5.2.3, be used to derive a limit for the number of packets in a connected subset of the intermediate nodes. This time we obtain an *upper* bound.

Theorem 5.5. *In any n consecutive intermediate nodes using our protocol, there are never more than $n + 1$ queued packets.*

Proof Let the joint queue state of the n nodes be denoted by $x_1/y_1, \ldots, x_n/y_n$. For each pair of consecutive nodes with states x_i/y_i, x_{i+1}/y_{i+1}, $1 \leq i < n$, there is, by the previous lemma, either $x_i = 0$ or $y_{i+1} = 0$. Furthermore, from the backpressure rule, we know that for all i with $1 \leq i \leq n$, both $x_i \leq 1$ and $y_i \leq 1$ hold. Therefore, $\forall i, 1 \leq i < n : x_i + y_{i+1} \leq 1$. For the total number of packets it thus holds that

$$
\begin{aligned}
\sum_{i=1}^{n}(x_i + y_i) &= y_1 + \sum_{i=1}^{n-1}(x_i + y_{i+1}) + x_n \\
&\leq 2 + \sum_{i=1}^{n-1}(x_i + y_{i+1}) \\
&\leq 2 + n - 1 \\
&= n + 1.
\end{aligned}
\tag{5.12}
$$

$\qquad\square$

Again, this can be used to obtain a bound on the total number of packets along the route.

Corollary 5.6. *For a bidirectional connection over h hops using out protocol, there are at most h packets in the network at any point in time. The number is temporarily undercut.*

Proof The first assertion follows directly from the previous theorem. The second part is rather obvious: when a packet leaves the network, the number of packets decreases, until a new packet is inserted. □

This demonstrates that the proposed distributed algorithm stays within the same bounds on the number of packets in the network as the centralized scheduler.

5.3.3 Dealing with Real Wireless Media

The astute reader will surely have noticed that, so far, we have assumed that all transmissions are always successful. In a real wireless network this is for sure not the case. Wireless interference can make practically every transmission fail. For a coded transmission, this may mean that one or both of the intended receivers cannot decode the message. Thus, we now look at how it is possible to recover from such a situation without sacrificing the desirable properties we have pointed out so far.

Since CXCC uses a similar backpressure rule as the one introduced above, integrating these approaches and adopting CXCC's single-hop reliability mechanism becomes a natural solution. It turns out that only one significant modification to CXCC is necessary to yield the desired behavior: adding the rule that only coded pairs of oppositely directed packets may be transmitted by intermediate nodes after leaving the initialization state.

One more aspect, however, also deserves attention. In noCoCo, when a node has made a transmission, it has to wait for feedback from both its neighbors before the next transmission is allowed. When a transmission fails, additional protocol handshakes are required for recovery. It is important to design these in a way that always ensures a consistent view of a neighbor's queue state. For example, it must not happen that a node A learns that its neighbor B has forwarded the previously sent packet (through an implicit or explicit ACK), without at the same time obtaining the information that it had been combined with a transmission in the opposite direction. This is particularly crucial during the initialization phase: otherwise the backpressure at A is released through the ACK, allowing A to forward another (uncoded) packet to B. It should, however, have sent a *coded* combination with the oppositely directed packet. Therefore, we need to make sure that we both allow recovery from packet losses *and* ensure consistent state updates at the neighboring nodes.

Fortunately, this is relatively simple to achieve. Data packets are coded and thus always carry information on both forwarding directions, from which the queue states can be inferred. Control packets can be tailored to contain information on a node's left- and right-directed queue states. In particular, a packet in one direction must never be acknowledged without also communicating the queue state of the opposite direction at the same time. This is achieved by also sending the respective control messages in pairs, one for each direction. In effect, each correctly received packet conveys information on both directions. The resulting protocol is able to effectively recover from packet losses, while retaining all the desirable properties shown previously.

5.3.4 Handling Finite Bursts of Data

As the final step to a practically usable protocol, we need to take into account that practical data transmissions are of finite length. In the protocol described so far, there is no way to return to a state where an intermediate node is allowed to continue with unidirectional transmissions, if the packet stream from one of the end nodes has ended, temporarily or permanently. Therefore, a mechanism is needed to allow for the completion of a transmission, in one or in both directions.

A viable solution is in fact pretty simple: when the source node sends the last packet of a burst, it may set a special flag in this packet. An intermediate node, after forwarding a packet with this *end-of-burst flag* set, will resume the initialization state. Unidirectional transmissions in the opposite direction are then allowed again at this node.

Just like with BarRel's actually quite similar LAST flag, the end of a burst may be indicated by the source application itself, if such a tight integration is desired and possible. Otherwise, a source node may simply set the flag if it sends out the last packet in its queue, i. e., if no further packets have so far been generated by the application.

In practice, this approach yields a smooth, load-dependent transition between opportunistic and enforced coding: if both sources produce packets at a rate that fills up the capacity of the route, maximum coding gain will be enforced. If at least one direction does currently not produce packets at a rate that suffices to build up backpressure, the scheme falls back to opportunistic coding. For unidirectional traffic, noCoCo reduces to plain CXCC, and therefore provides efficient congestion control for the unidirectional flow.

5.4 Performance Evaluation

In order to assess the performance of co-ordinated coding against a purely opportunistic approach, we have performed simulations using the network simulator ns-2.30 [ns2a]. For performance comparison, we have implemented three protocols:

- COPE, as described in [KRH+06]; just as UDP/TCP over 802.11 and CXCC, it serves as a reference. A simulator implementation of COPE was previously not available.

- A straightforward combination of CXCC and COPE-style opportunistic network coding; this protocol performs CXCC-style packet forwarding, implicit acknowledgments, etc., but combines multiple packets via XOR into one transmission if coding opportunities arise. The existence of such coding opportunities is, however, not guaranteed.

- The co-ordinated network coding scheme we introduce.

In all but our most simple simulation scenarios there will be more than one connection, and thus additional coding opportunities may arise. The COPE-based protocols depend on recognizing these. Our noCoCo implementation also makes use of coding opportunities beyond the current two-way connection, where they arise and can be identified.

However, COPE's reception report and guessing mechanisms both incur substantial complexity and add variance to the protocol performance. Given our focus is on the different performance due to co-ordinated versus opportunistic coding, we establish upper and lower bounds on what *any* mechanism for identifying coding opportunities can possibly achieve, instead of following the above mechanisms.

For this purpose, we derive two variants for each coding scheme, termed "conservative coding" and "omniscient coding". Conservative coding combines packets only if decoding is guaranteed successful, without exchanging any additional information. In effect, overheard packets are not used for coding. Conservative coding thus represents a lower bound of what can be achieved by any scheme for identifying coding opportunities. Omniscient coding, on the other hand, employs a central component to provide each node with immediate and exact information on the packets known by other nodes. While this can be implemented in a simulator, it is obviously not possible on real devices. Omniscient coding allows for perfect coding decisions, without the delay or overhead of reception reports or the risk of guessing wrong. This gives an upper bound on what an ideal scheme may achieve for identifying coding opportunities. We will show later that the lines representing either bound often overlap for noCoCo in the plots.

We use the same general simulation parameters that we have also employed before. This includes a physical layer bit rate of 1 MBit/s, the two-ray ground propagation model, 250 m radio range and 550 m carrier sense radius. Data packets carry 512 bytes of payload. We use static, hop-count minimal routes to avoid possible side-effects introduced by a specific routing approach. The RTS/CTS mechanism of 802.11 was again switched off in all our simulations. In the following, we first investigate the protocols' performance in simple chain and cross topologies like the ones we have used for CXCC in Chapter 2, and then consider random scenarios with dynamic traffic patterns.

5.4.1 Chain Topology

For our first set of simulations, we use a chain topology with ten hops. The distance between neighboring nodes is 150 m. We set up bidirectional UDP traffic, originating from both ends of the chain, and being directed to the respective opposite end. The offered load at the sources is gradually increased.

All intermediate nodes can encode at most two packets together at a time, one from each direction. In such an environment, there is no difference between the coding opportunities that can be identified with conservative coding and with omniscient coding. Therefore we do not need to distinguish between these two.

Figure 5.3 shows how the total application-layer throughput varies with increasing offered load. Opportunistic network coding alone barely alleviates the rapid performance deterioration of bidirectional UDP traffic once the optimal load is exceeded. But coordinated network coding with noCoCo achieves superior throughput. One reason is the implicit backpressure property of the congestion control mechanism, which is also present in CXCC. But noCoCo achieves a substantially higher throughput than CXCC, and this throughput gain far exceeds the one obtained with opportunistic coding in CXCC+COPE. The main reason for the smaller gain with the opportunistic approach is that coding opportunities do not reliably arise. Nodes often have only one packet in the transmission queue, and many transmissions are uncoded.

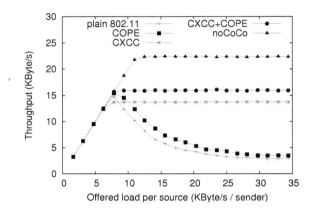

Figure 5.3: Throughput in chain topology.

Figure 5.4 shows the average packet delay, measured from when a packet leaves the source node to its successful reception at the destination. The delay increases very quickly with increasing offered load for plain 802.11 as well as for COPE, because long queues are building up, especially in the nodes close to the ends of the chain. A packet has to wait for a potentially long time in these queues, before being forwarded. Since the backpressure rules in noCoCo and CXCC result in very short queues in the intermediate nodes, the delays are substantially shorter. As a consequence of even better medium utilization, noCoCo's delays are smaller than those of opportunistic coding schemes.

The efficiency of the medium utilization is also closely related to the protocol overhead. We again use the overhead metric that quantifies the average amount of data transmitted on the wireless medium in order to bring one byte of payload one hop further. Figure 5.5 presents the results of this evaluation in the chain topology simulations. Apart from generally confirming the picture gained from the previous metrics, the noCoCo plot impressively demonstrates the benefits of network coding. Around an offered load of 10 KB/s at each source the source data rate approaches the network capacity. At this point, the transition from opportunistic forwarding of single packets to enforced coding happens for noCoCo.

Without network coding, the optimal value of our overhead metric is one: it is clearly not possible to forward one byte of payload while transmitting less. Once network coding at each hop is guaranteed, noCoCo underruns the value of one for the overhead metric: on average it transmits about 0.79 bytes to forward one byte of payload over

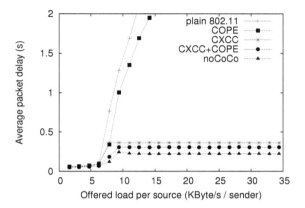

Figure 5.4: Packet latency in chain topology.

one hop. COPE does not achieve this, again due to the lack of spontaneous coding opportunities.

5.4.2 Cross Topologies

We now turn to cross topologies, consisting of two orthogonally aligned chains, sharing one node in the middle. UDP traffic flows from the end of either leg of the cross to the end of the opposite leg. We first increase the offered load in a cross with five hops in each leg, similar to what we did in the above chain simulations. Subsequently we will look at the effects of varying leg lengths.

Due to the node spacing of 150 m and the communication radius of 250 m, the nodes adjacent to the center node are able to overhear the transmissions of their two counterparts in the other chain. Thus, it is generally possible to combine up to four packets into one transmission at the middle node. While omniscient coding can make optimal use of this, conservative coding will not. We therefore distinguish between these two strategies in our figures, denoting conservative coding by "(c)" and omniscient coding by "(o)".

Figures 5.6, 5.7, and 5.8 show throughput, packet delay, and overhead respectively for the cross with a leg length of five hops. The occurring effects are generally quite similar to those already observed in the chain topology. Remarkably, there is generally very small difference between conservative coding and omniscient coding. This suggests

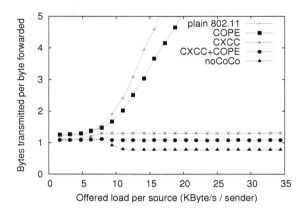

Figure 5.5: Overhead in chain topology.

that the benefit from additional coding opportunities is quite limited. For noCoCo in particular, the differences between the simulations with conservative coding and with omniscient coding are so small that the respective lines in the plots are barely distinguishable.

One might argue that these small differences stem mainly from having only 1 out of a total of 21 nodes that is able to make use of the additional opportunities. We thus complement the results with plots showing the effects of varying leg lengths of the cross. In Figures 5.9, 5.10, and 5.11 we use a saturated offered load and gradually increase the size of the cross. From these results, it becomes clear that the previously discussed overall picture of relative performance sets in very quickly, and generally holds for a leg length of two hops already.

The only significant differences occur for the situation where each leg is only one hop long, i.e., where the source/sink nodes are directly adjacent to the center node. In this case, the differences between conservative and omniscient coding are indeed significant, particularly for COPE. The effect of a high gain of COPE versus plain 802.11 in this specific setting has been observed and explained as the "coding+MAC gain" in [KRH+06]. Without coding the intermediate node needs to transmit four times as often as the other nodes, but the 802.11 MAC does only assign it 1/5 of the medium time. With omniscient coding, this limitation is no longer a bottleneck. The central node will not be able to access the medium more often, but it can transmit up to four packets with one medium access.

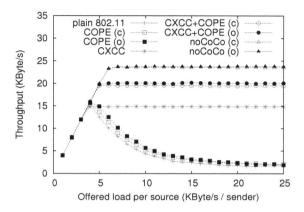

Figure 5.6: Throughput in cross topology with increasing offered load.

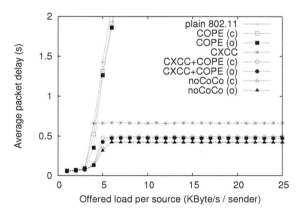

Figure 5.7: Packet latency in cross topology with increasing offered load.

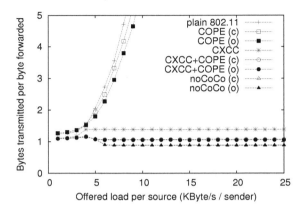

Figure 5.8: Overhead in cross topology with increasing offered load.

The backpressure rules in noCoCo and the CXCC-based schemes make the source nodes refrain from further transmissions until the central node has forwarded the previous one. These protocols therefore inherently avoid the problem of inappropriate assignment of medium access opportunities.

5.4.3 Random Topologies

Finally, we study gains obtainable with noCoCo versus opportunistic schemes in more practical settings, and consider random topologies. We intentionally set them up in a way that yields rapidly changing traffic patterns. With few, long-lived connections the availability of coding opportunities essentially depends on the routes of these connections and whether they share many intermediate nodes, whereas many short-lived will result in a large variation in the packets that meet.

Each simulation scenario uses 150 nodes at uniformly random positions on a $1500 \times 1500\,\text{m}$ square area. A total of 40 bidirectional connections start at random times between 0 and 120 seconds. Each connection is assigned a random amount of data between 5 and $50\,\text{KB}$, which is to be transmitted in both directions. In the absence of route breaks the single-hop reliability mechanism of CXCC will result in reliable end-to-end delivery; the same holds for noCoCo, which adopts CXCC's respective mechanisms. It does not hold for plain 802.11 and COPE. Thus, in order to ensure reliable delivery of the data we use TCP Newreno as a reliable transport protocol. Of course, the problems of TCP congestion control over wireless multihop networks have to be taken

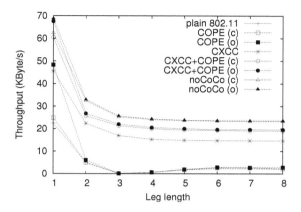

Figure 5.9: Throughput in cross topology with increasing leg length.

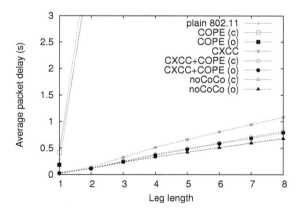

Figure 5.10: Packet latency in cross topology with increasing leg length.

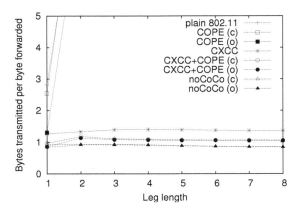

Figure 5.11: Overhead in cross topology with increasing leg length.

into account when comparing the results here. However, since we are more interested in the relative improvements with opportunistic and co-ordinated network coding, this is only of secondary relevance.

We simulate all protocols using the same set of topologies and traffic patterns. We calculate the throughput of each single connection, by dividing the amount of data delivered by the time it took from initiating the connection to the successful delivery of the last data segment. In Figure 5.12 we show the cumulative distribution functions of these per-flow throughputs. For example, about 25 % of all connections with noCoCo achieve a throughput below 10 KByte/s, whereas the same applies to about 40 % of the connections with CXCC, and to about 70 % of those with TCP Newreno over IEEE 802.11.

The results with conservative coding and omniscient coding are generally close together. To maintain readability of the figure, we only show the results with omniscient coding for COPE and CXCC+COPE (as an upper bound of these protocols' performance), and with conservative coding for noCoCo (as a lower bound).

The random topology simulations confirm that our findings from the deterministic topology simulations above hold in a similar way also in more complex environments. It was pointed out in [KRH+06] that the interaction between COPE and TCP is complex, due to TCP's congestion avoidance rules, timing issues and potential packet reordering. We also make this observation here. Even though the traffic is bidirectional—and coding opportunities of similarly sized packets can thus generally exist at each single intermedi-

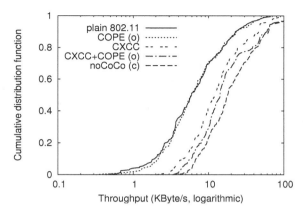

Figure 5.12: Cumulative distribution function of throughputs in random topologies.

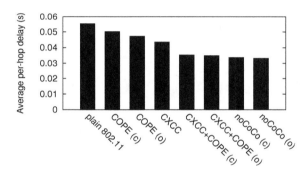

Figure 5.13: Per-hop delay in random topologies.

ate hop—COPE barely improves the throughput. For the protocols with a backpressure rule, the relative performance matches the picture from the previous simulations, with opportunistic coding noticeably improving upon the performance of non-coding CXCC, but in turn being clearly outperformed by co-ordinated network coding.

The average per-hop delay of the connections is shown in Figure 5.13. The relative ordering of the protocols remains the one observed above, though the absolute differences are generally not as grave as those of, e. g., the throughput.

In Figure 5.14 we have evaluated the overhead in the random topology simulations, and again the overall picture seems familiar. In the presence of a complex traffic pattern and with TCP being used, COPE actually exhibits a minimally higher overhead than plain

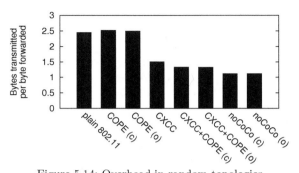

Figure 5.14: Overhead in random topologies.

802.11. This can be traced back to a higher number of packet retransmissions. Unlike in the chain and cross topologies above, noCoCo does not underrun the non-coding limit of an overhead of one in these simulations, but it comes very close to this value.

5.5 Chapter Summary

In this chapter, we have introduced a deterministic scheme for network coding within two-way traffic flows in wireless multihop networks. We have derived some general properties of scheduling schemes that achieve maximum coding gain, and we have introduced a centralized scheduling approach with favorable performance. Developing this further into a practical and distributed protocol, we have proposed Near-Optimal Co-ordinated Coding (noCoCo). It is based on implicit co-ordination and implicit backpressure, using rules that can be understood as an extension of CXCC. noCoCo has been shown to approach the theoretical limits established with the centralized scheduling approach.

Our evaluation demonstrates the potential of a co-ordinated approach for network coding, in terms of throughput, packet delay, and protocol efficiency. The results also reveal some interesting insights regarding opportunistic coding schemes. Without appropriate medium allocation, network coding alone may also be at its wit's end for achieving good performance. It is also interesting to notice the small differences between conservative coding and omniscient coding. This suggests that the benefits from identifying additional coding opportunities are limited in this context, which could be used to simplify coding protocols adopting similar scheduling principles.

Chapter 6

Post-Facto Offline Time Synchronization

In the preceding chapters of this thesis, we have considered alternative protocol designs for wireless multihop networks. The introduced approaches CXCC, BarRel, BMCC, and noCoCo make use of the medium's specific properties, while the network is in operation. For example, the local broadcast property allowed for implicit acknowledgments and backpressure. The fact that transmissions are locally serialized made very short queues possible, and, as a consequence, implicit end-to-end reliability practicable. In this chapter, we will now demonstrate that the concept reaches beyond applications in network protocols: well-directed use of implicitly obtained information can also help to overcome other difficulties.

A fundamental problem with interpreting results from real-world experiments in computer networks is that each system uses its own local clock to timestamp events. These clocks do not run perfectly synchronous, they can deviate. At the end of the experiment, the result is a set of log files where the timestamps are based on the clocks of the individual systems. This is generally insufficient for the investigation of timing issues and the correlation of events. It is thus highly desirable to obtain an event log where all timestamps refer to a single reference clock instead of multiple local clocks.

The obvious attempt to reach this goal is to synchronize the clocks of the involved systems with a time synchronization protocol like NTP [Mil94, Mil92]. Unfortunately, there are two key reasons why this approach may be inappropriate. First, running a time synchronization protocol will cause additional network traffic and may thus interfere with the experiment. Second, even if the clocks were perfectly synchronized, it takes some system dependent (and potentially non-deterministic) time from the occurrence of an event until it is actually timestamped and recorded. We call this the *timestamping delay*. The drawback of additional network traffic can be eliminated by providing each system with a very accurate clock, e.g., controlled by a GPS receiver. Aside from the fact that this is quite expensive, this approach does not solve the problem of the timestamping delay. While it may also be possible to use customized hard- and software to bound the timestamping delay, such a solution cannot be employed for the off-the-shelf systems often used in network experiments.

In order to avoid these problems we propose to correct the timestamps of the individual log files after an experiment is completed instead of synchronizing clocks during the experiment. To do so, we again exploit the local broadcast property: a transmission is often observed by multiple nodes at the same time. If the same event has at the same time instant been observed and locally timestamped by more than one system, it can be used as an anchor point for a post-experiment synchronization of the log files. These anchor points can be extracted from the set of locally written log files after the experiment. Thus, our scheme utilizes the communication that occurs anyway during the experiment, and does not require exchanging any further data between the nodes. Therefore, such an approach to time synchronization is a prime example for the use of implicitly obtained knowledge.

In essence we employ a model for the clocks and the timestamping delays. We then use the anchor points to estimate the parameters of this model and thus the clock deviations. This results in estimates for the timestamps of all events on a common time basis. A maximum likelihood estimator is used. It leads to a large linear program with a very specific structure. We exploit this structure to solve the linear problem efficiently in spite of its huge size. The solution then yields a synchronized log file where all entries are recorded with a common time basis.

Analytical and numerical results show that the solution converges quickly to a good estimate for increasing input data sizes, and that it is robust if the assumptions made for its derivation are not perfectly fulfilled. Thus, in practice, a very reasonable amount of log data is typically sufficient to identify and eliminate clock deviations to a very large extent. It thus allows for an in-depth analysis of the events in an experiment. Beyond the extraction of timing relations from the experimental data, it could, for instance, also support the fine-grained visualization of experiments, as our tool Huginn—presented in [SFT+05a, SFT+05b, Sch07]—did for wireless multihop network simulations.

The presented time synchronization approach is not only applicable to wireless multihop networks, but also to other networks with local broadcast characteristics. It just requires that the clocks of any two nodes in the network can be—directly or indirectly—set into relation by anchor points. In particular this includes experiments in wireless ad-hoc, sensor, and mesh networks, as well as local area networks with multiple stations in each collision domain and satellite networks. A paper covering the central results from this chapter is currently under review [SKR+b].

6.1 Related Work

The relevant literature in the area of clock synchronization can be divided into online and offline clock synchronization protocols. The aim of *online clock synchronization* protocols, like the well-known Network Time Protocol (NTP) [Mil94, Mil92], is to keep the clocks of the participating nodes synchronized while the network is up and running. In contrast to that, *offline clock synchronization* approaches correct timestamps that have been provided by unsynchronized clocks after the experiment is finished. Our own approach clearly falls into the second category.

6.1.1 Online Clock Synchronization

As discussed in the introduction, online approaches typically use explicit messages for clock synchronization. They are also constrained by the fact that they need to work in a distributed fashion and may consume only very limited computational resources. Moreover, online synchronization can only exploit past information, whereas offline approaches can make use of all—previous as well as later occurring—events for the time estimates. For these reasons, online synchronization protocols are not an optimal solution for the synchronization of distributed log files. Nevertheless, some of them use the idea of events that are observed by multiple systems. In the following, we summarize those approaches. A broader overview of the topic, with a focus on wireless sensor networks, can be found in [RBM05].

A number of online synchronization protocols [VRC97, MFNT00, EGE02] relies on the parallel reception of broadcasted packets by multiple systems. A broadcasted packet is received by all systems nearly at the same instant, the only uncertainty in timestamping such packets is the signal propagation time and the timestamping delay. To synchronize the clocks, the recipients of a given broadcast communicate to exchange their respective reception times. By comparing these reception times, two nodes are able to compare and adjust their clocks. In [EGE02], for example, the clock skew is estimated using linear least squares regression. A complete network can then be synchronized by synchronizing adjacent nodes pairwise along a tree structure, yielding, however, the disadvantage of accumulating the pairwise errors.

In [KEPS04], the pairwise synchronization of [EGE02] is extended to a global one. The authors present an online synchronization approach for sensor networks that is based on a global unbiased minimum variance estimator. They first introduce a version that considers only clock offsets, and then complement it with an idea on how to deal with clock rate differences. Their approach is, however, not able to handle offsets and rate deviations conjointly, but must rely on separate estimates on different time scales. This is feasible and appropriate in the considered context of online time synchronization for continuously running sensor networks, but is not optimal for the offline synchronization of the logs of time-limited experiments. In addition to avoiding the general drawbacks of using online approaches for the synchronization of log files, our approach estimates offsets and rates in one single step, and can thus exploit all the available information to find the global optimum for both.

6.1.2 Offline Clock Synchronization

The first offline clock synchronization algorithm has been proposed by Duda et al. [DHHB87] for generic distributed systems. The send and receive timestamps of messages between nodes A and B are taken as coordinates of a point, the x-axis being the timestamp of A and the y-axis being the timestamp of B for the same packet. Due to the network delay, two point-clouds emerge with an empty corridor in between. Each point is either above the corridor (when sent from A to B) or below (when sent from B to A). The authors present two methods to fit a line in this corridor, thereby estimating the difference in clock speed and offset between A and B. The first method computes the separating line with linear regression, the other uses a convex hull approach. They also sketch a maximum likelihood approach but are not able to use it due to a lack of knowledge about the message delay from sender to receiver, which would be needed.

Duda's linear regression and convex hull approaches have been extended in [Ash95]. The author corrects the timestamps using experimental knowledge about the smallest round trip delays. This knowledge is incorporated in an algorithm that selects the two best points to estimate the skew and offset between the nodes. In [MST99], linear programming is used to compensate for clock skew that influences one-way delay measurements between two nodes over the Internet. A convex hull based approach able to cope with clock resets is presented in [ZLX02].

All of the presented offline synchronization algorithms can compensate linear clock deviations between two nodes without requiring additional network traffic. In contrast to our approach, which exploits the broadcast nature of the medium, they can be used for all kinds of communication systems. However, this benefit is also their main drawback: all of them consider the comparison of send and receive timestamps. Thus, the network delay cannot be completely eliminated, as it is the case in our approach. Likewise, they cannot separate and handle the timestamping delay. Finally, while we use all the available data to compute globally consistent estimates for an arbitrary number of nodes in parallel, all these algorithms synchronize only two clocks directly. In order to synchronize more clocks, a successive synchronization of node pairs is necessary, a process in which errors can accumulate.

6.2 Model and Terminology

6.2.1 Nodes and Events

In our terminology, an *event* is an incident that has been observed by one or more nodes and is recorded in their local log files. Of particular interest for us are packet reception events, since they can be observed by multiple nodes almost at the same time. Broadcast packets can generally be received by all nearby nodes, for unicast transmissions a similar effect can be achieved by logging receptions in promiscuous mode. In this mode, a node records all receptions that could be decoded by its network interface, regardless of whether the node is the intended destination of the transmission or not. We assume that parallel receptions of the same transmission can be identified as such. In the following, we concentrate primarily on events that have occurred in more than one node, since they can serve as anchor points for the synchronization.

We denote the set of nodes participating in an experiment by J and the set of events that occur during the experiment by I. Each event $i \in I$ occurs at some "true" time T_i. The same event i can be observed by multiple nodes. In this case each of the nodes records its own timestamp for the event, according to its local clock, i.e., event i is recorded by some of the nodes $j \in J$ with local timestamps $t_{i,j}$.

The recorded times define a relation $R \subseteq I \times J$ in the sense that $(i, j) \in R$ if and only if event i is recorded by node j. The subset of nodes that receive a certain event $i \in I$ is denoted by R_i, i.e., $j \in R_i$ if and only if $(i, j) \in R$. Similarly, R^j stands for the set of events observed by node j.

6.2.2 Clocks

We model clocks as twice differentiable functions, mapping some (virtual) global, absolute time to the view of the respective clock. This model matches those commonly used in literature related to clocks and time synchronization [Mil92, MST99, EGE02], and is justified since only the limited timespan of a single experiment needs to be considered. For the same reason and for the sake of simplicity we do not account for clock resets, although our approach could be extended to such a scenario.

The *true clock* C_T is a clock which is correct by definition: $\forall t : C_T(t) = t$. Our aim is to approximate this clock as closely as possible by the calculated global event timestamps.

We use the common terminology to describe the properties of clocks. The *offset* of a clock C at time t is the difference $C(t) - C_T(t)$ between C and the true clock C_T. If we use the term offset without referring to a certain point in time we refer to $C(0)$, the offset at time $t = 0$. $C'(t)$ is called the *rate* or *frequency* of C at time t. The difference between a clock's rate and the true clock's rate $C'(t) - C_T'(t) = C'(t) - 1$ is called *skew*. Finally, the second derivative $C''(t)$ is called the *drift* of C.

Let $E \subset \mathbb{R}$ be the interval denoting the "real" timespan of an experiment. The main reason for clock drift are changes in the environmental conditions, mostly temperature changes, influencing the quartz oscillator of the clock. In [KM07], the MANET experiments described in the literature are analyzed; there, it has been pointed out that typically the duration of an experimental run does not exceed 1000 seconds. This coincides with previous work on clock stability, which shows that the clock drift is negligible over time spans up to 1000 seconds [VBP04]. It can thus reasonably be expected that the clocks' drift during E is negligible.

Consequently, our model assumes that, in E, the local clocks in the nodes can be closely approximated by a linear function. We denote the rate of the local clock C_j of a node j by $r_j > 0$ and its offset at time $t = 0$ by o_j. Thus we have

$$\forall t \in E : C_j(t) = r_j t + o_j. \tag{6.1}$$

For longer experiments, if the assumptions are not fulfilled, the accuracy of the results may deteriorate. Note, however, that it is easily possible to synchronize longer logs interval-wise, such that the linearity assumption holds reasonably well within each interval.

In practice, time in computer systems does not run continuously, but progresses in discrete steps. While the resolution of the timer-interrupt driven system clock is typically relatively coarse—in the order of milliseconds—, more fine-grained time sources are often available and used. On the x86 platform, for example, the CPU's TSC register progresses with every CPU clock cycle. Thus, its granularity is very fine. It serves for generating the timestamps, for example, when using a Linux kernel and the widespread packet tracing library libpcap. Thus, we can assume that the error introduced by the clock resolution is small in comparison to other sources of error. Our approach does not amplify such errors.

When performing experiments, it must be taken care that the linearity assumption is not thwarted by processes running on the nodes. In particular, no online time synchronization should be running in parallel. If online synchronization must be used—e.g., because it is part of the experiment—then it should record all the modifications it made to the local clock, so that the effects of these changes can be eliminated from the log files of the respective nodes prior to synchronizing them.

6.2.3 Timestamping Delay

When sending a message, a number of different delays occur from the moment the source application generates the message until the receiver timestamps it. As our approach uses these timestamps as synchronization anchors, we are interested in the delay *differences* experienced by distinct nodes. The deterministic components are not an issue in our context: if all timestamps in a node are recorded late by some fixed time, then this is the same as if they were recorded immediately with a correspondingly increased offset. So, the fixed delay components are equivalent to an additional clock offset.

The experienced delay can be decomposed into four components according to [KO87]: the time needed to compose the message and to assemble the packet, the time to access the medium, the propagation delay on the medium, and finally, after the transmission arrives at the receiver, the *receive time*, i. e., the delay for checking the message and recording the arrival timestamp. Obviously, the time until the packet leaves the sender is the same for all receivers and thus does not need to be considered.

The propagation delay depends on the distance between sender and receiver, and the propagation delay differences depend on the different distances between sender and receivers. As long as these differences are in the order of a few hundred meters, the propagation delay difference is in the order of at most microseconds and is therefore negligible[1]. The receive time occurs as the recording of the timestamps in the nodes does not happen immediately upon reception, but is delayed. It can be decomposed further into a fixed component (which equals the minimum path delay of the processing necessary at the receiver), and an additional, variable time that occurs because the timestamping is performed by the node's CPU, which may be busy with other tasks before the event is processed. The latter we call *timestamping delay.*

Note that the delay of an event is also "measured" by the recording node's clock, and thus is scaled with the rate of this clock. A delay $d_{i,j}$ thus leads to a timestamp

$$t_{i,j} = C_j(T_i + d_{i,j}) = r_j(T_i + d_{i,j}) + o_j. \tag{6.2}$$

The timestamping delay is, like all delays, obviously nonnegative. Furthermore, it seems reasonable to assume that most timestamps are recorded with small latency and few are set after a longer time. We model the timestamping delays as exponentially distributed, pairwise independent random variables. Moreover, we assume that the exponential distributions of all delays share the same parameter λ. The latter is reasonable if the nodes participating in the experiment use comparable hard- and software for the timestamp generation, which will often be the case in a testbed.

In a real-world application our assumptions about the timestamping delay just like that about the clocks' linearity will of course not perfectly hold. In fact, depending on the hard- and software of the devices, reality might look very different. We use the mentioned assumptions for the motivation and derivation of our method. It will later become clear that the resulting approach yields good results also under non-conformant circumstances.

6.2.4 Connectivity Constraints

As approach relies on anchor points to set the clocks of nodes into relation, it depends on the availability of such anchor points, and on them setting all clocks in the network into relation. Observe that synchronization on the basis of anchor points is impossible otherwise: if there is no common event between two groups of nodes, it would be impossible to tell if all the clocks in one of the groups are, for example, early by one hour. Thus, a common time basis cannot be established. Such a situation is in fact not

---

[1]If nodes are really far apart and the propagation delay is long, then it is often the case that the distances and thus also the delays are approximately known. This applies, e. g., to satellite systems. In this case it is possible to eliminate the delay prior to synchronization.

different from two completely unrelated experiments. However, it of course still remains possible to synchronize the clocks *within* each group.

Note that the availability of anchor points does *not* imply that all pairs of nodes must share common events—clocks may also be related indirectly, over intermediate nodes. It also does not necessitate that the network is "connected" in the commonly used sense. For example, if there are two almost independent groups of nodes, one single node sharing events with both groups suffices. These shared events need not occur during the same time intervals, and thus there is no need for network connectivity at even just one single point in time, as long as nodes move between partitions.

Hence, the existence of anchor points is not a very hard constraint in practice, and anchor point-based synchronization will be possible in the vast majority of experimental setups. Otherwise, it is still possible to apply our scheme, if artificial anchor points are generated, e. g., by broadcasting "anchor packets". Doing so during the experiment brings one of the drawbacks of online synchronization with it—possible interference with the experiment itself. Anchor points may, however, also be generated before and after an experimental run.

6.3 Algorithm

The previous section has introduced a model of the network and the timestamping delays. Now, we will formalize the problem and propose an approach for its solution via a maximum likelihood estimator (MLE). Given the recorded local timestamps, we wish to maximize the likelihood that our estimates of the true event times are correct.

Due to the exponentially distributed delays, the conditional probability density for measuring a timestamp $t_{i,j}$ for event i at node j given C_j and T_i is

$$f(t_{i,j} \mid C_j, T_i) = f(d_{i,j}) = \lambda e^{-\lambda d_{i,j}}. \qquad (6.3)$$

Because of the independence of the delays the probability density for the whole set of measurements in our experiment can be written as

$$f((t_{i,j})_{(i,j)\in R} \mid (C_j)_{j\in J}, (T_i)_{i\in I}) = \prod_{(i,j)\in R} \lambda e^{-\lambda d_{i,j}}. \qquad (6.4)$$

We can now express our problem as an optimization problem. Under a uniform prior, we want to find the optimal estimates $\widehat{T}_i$ of T_i for all $i \in I$, and, in parallel, the optimal estimates $\widehat{C}_j$ of C_j for all $j \in J$ such that the likelihood function L defined in the following way is maximized:

$$
\begin{aligned}
L &= L((\widehat{C}_j)_{j \in J}, (\widehat{T}_i)_{i \in I} \mid (t_{i,j})_{(i,j) \in R}) \\
&= f((t_{i,j})_{(i,j) \in R} \mid (\widehat{C}_j)_{j \in J}, (\widehat{T}_i)_{i \in I}).
\end{aligned}
\tag{6.5}
$$

From (6.2) we can see that

$$
\forall (i,j) \in R : d_{i,j} = \frac{t_{i,j} - o_j}{r_j} - T_i.
\tag{6.6}
$$

This relation must also hold for the estimates of T_i and C_j. Let $\widehat{r}_j$, $\widehat{o}_j$, and $\widehat{d}_{i,j}$ denote the estimates for r_j, o_j, and $d_{i,j}$, respectively. Then, in analogy to the above we have

$$
\forall (i,j) \in R : \widehat{d}_{i,j} = \frac{t_{i,j} - \widehat{o}_j}{\widehat{r}_j} - \widehat{T}_i.
\tag{6.7}
$$

Therefore, L can be expressed as

$$
\begin{aligned}
L &= \prod_{(i,j) \in R} \lambda e^{-\lambda \widehat{d}_{i,j}} \\
&= \prod_{(i,j) \in R} \lambda e^{-\lambda \left(\frac{t_{i,j} - \widehat{o}_j}{\widehat{r}_j} - \widehat{T}_i \right)},
\end{aligned}
\tag{6.8}
$$

eliminating the estimates $\widehat{d}_{i,j}$ for the unknown quantities $d_{i,j}$.

Since all the delays are non-negative, the maximization of L is subject to the constraints

$$
\forall (i,j) \in R : \frac{t_{i,j} - \widehat{o}_j}{\widehat{r}_j} - \widehat{T}_i \geq 0.
\tag{6.9}
$$

Now we apply a standard technique in maximum likelihood estimation: maximizing L

is equivalent to maximizing $\ln L$, because $L > 0$ for all valid parametrizations.

$$\ln L = \ln \prod_{(i,j) \in R} \lambda e^{-\lambda \left(\frac{t_{i,j} - \widehat{o}_j}{\widehat{r}_j} - \widehat{T}_i \right)}$$

$$= \sum_{(i,j) \in R} \left(\ln \lambda + \ln e^{-\lambda \left(\frac{t_{i,j} - \widehat{o}_j}{\widehat{r}_j} - \widehat{T}_i \right)} \right) \tag{6.10}$$

$$= |R| \ln \lambda - \sum_{(i,j) \in R} \lambda \left(\frac{t_{i,j} - \widehat{o}_j}{\widehat{r}_j} - \widehat{T}_i \right).$$

Optimizing this expression with regard to λ and all the $\widehat{T}_i$ and $\widehat{C}_j$ is a difficult nonlinear optimization problem. However, we are not primarily interested in the parameter λ. Fortunately it turns out that the optimal $\widehat{T}_i$ and $\widehat{C}_j$ are independent of the value of λ. Let for the moment

$$k(x) := -\frac{\ln x - |R| \ln \lambda}{\lambda}. \tag{6.11}$$

k is strictly monotonically decreasing for any $\lambda > 0$ and $|R|$. Thus, it is easy to see that L is maximal if and only if $k(L)$ is minimal:

$$k(L) = -\frac{\ln L - |R| \ln \lambda}{\lambda} = \sum_{(i,j) \in R} \left(\frac{t_{i,j} - \widehat{o}_j}{\widehat{r}_j} - \widehat{T}_i \right). \tag{6.12}$$

Therefore, instead of maximizing L, we minimize $k(L)$. We have thus eliminated the variable $\lambda > 0$. The constraints of the resulting optimization problem are still of the form (6.9).

From the clock model we know that the rates of the clocks are strictly positive. We exploit this fact and define

$$\bar{r}_j \quad := \quad \widehat{r}_j^{-1} \tag{6.13}$$

$$\bar{o}_j \quad := \quad \frac{\widehat{o}_j}{\widehat{r}_j}. \tag{6.14}$$

Equivalently, we have $\widehat{r}_j = \bar{r}_j^{-1}$ and $\widehat{o}_j = \bar{o}_j \widehat{r}_j = \frac{\bar{o}_j}{\bar{r}_j}$. Expressing $k(L)$ in terms of the variables $\bar{o}_j$ and $\bar{r}_j$ leads to

$$k(L) = \sum_{(i,j) \in R} \left(t_{i,j} \bar{r}_j - \bar{o}_j - \widehat{T}_i \right). \tag{6.15}$$

Similarly, the constraints (6.9) can be simplified to

$$\forall (i,j) \in R : t_{i,j}\bar{r}_j - \bar{o}_j - \widehat{T}_i \geq 0. \qquad (6.16)$$

This is a linear objective function with linear constraints, which can be solved using standard linear program (LP) solvers like, e. g., the simplex method.

For exponentially distributed errors, the maximum likelihood estimator is known to be nearly optimal. In our case, however, a different interpretation of the resulting approach is also possible. When comparing (6.12) and (6.7), we observe that

$$k(L) = \sum_{(i,j) \in R} \widehat{d}_{i,j}. \qquad (6.17)$$

The optimal solution minimizes the sum of the estimated delays. Therefore, the resulting approach may also be understood as a form of constrained Least Absolute Deviation (LAD) regression. Since this interpretation is completely independent from the assumption of exponentially distributed delays, it supports the expectation that the derived estimator is also well-suited for delays with other distributions.

Note that the optimization problem (6.15) and (6.16) has the trivial solution $\forall j \in J :$ $\bar{o}_j = \bar{r}_j = 0$ and $\forall i \in I : \widehat{T}_i = 0$. This is because, from the information contained in the log files, it is not possible to estimate all the absolute rates, but only the relative deviation between clocks. We call this the *rate ambiguity*. To overcome the rate ambiguity, we add a normalizing constraint $\sum_{j \in J} \bar{r}_j = |J|$; in the average, the inverse clock rates are assumed to be accurate. This assumption, however, is not crucial at all: if the average takes some other value, the solutions are simply scaled accordingly.

Similar to the rate ambiguity, there is also an *offset ambiguity* in the log files. The right hand sides of (6.15) and (6.16) do not change when all $\bar{o}_j$ are replaced with $\bar{o}_j + \tau$ and all $\widehat{T}_i$ are replaced with $\widehat{T}_i - \tau$, where $\tau \in \mathbb{R}$ is a given constant term. Thus, like above for the rates, it is not possible to estimate absolute, but only relative event times and clock offsets (even ignoring the fact that there is, of course, no "absolute time"). We may set, without loss of generality, $\bar{o}_1 = 0$.

If a reference clock is available—e. g., because at least one node has a connection to an external time source like a GPS receiver and records appropriate data—absolute synchronization to this reference is possible. More specifically, if the correct, global time of one event occurrence in one single node is known, then the offset ambiguity can be overcome. If the global times of any two events, or, alternatively, the time of one event and the rate of one node are known, then the rate ambiguity can likewise be eliminated. This is possible either by adapting the constraints for rates and offset, or by a respective transformation of the synchronization result.

The resulting linear program can be written in the form

$$\text{minimize } b^\mathsf{T} y \quad \text{subject to} \quad A^\mathsf{T} y \leq c, \tag{6.18}$$

where y is the vector of the unknowns $\widehat{T}_i$ for $i \in I$, followed by the vectors $\bar{o} \in \mathbb{R}^{|J|}$ and $\bar{r} \in \mathbb{R}^{|J|}$ of the $\bar{o}_j$ and $\bar{r}_j$ for $j \in J$, i.e.,

$$y = \begin{pmatrix} \widehat{T} \\ \bar{o} \\ \bar{r} \end{pmatrix}. \tag{6.19}$$

The matrix A^T represents the inequality constraints (6.16) and the normalizing constraints.

Events that have only been observed by one single node do not contribute information for the synchronization. Therefore, to keep the size of the linear program as small as possible, they should not be included in the optimization. Corrected timestamps for such events can, however, easily be generated based on the rate and offset estimates.

6.4 Solving the Optimization Problem

In (6.18), (6.19) the maximum likelihood estimator is defined as the solution of a linear program with $|I| + 2 \cdot |J|$ variables and $|R|$ linear inequality constraints. Due to the size of the linear program a straightforward application of the simplex method may result in a significant effort in terms of computational power and memory. When solving (6.18) with a standard simplex solver like QSopt [ACDM] the program takes hours to terminate even for relatively small problems. Therefore, we will now focus on the special structure of the linear program (6.18) and how it can be exploited to allow for a fast numerical solution. Below we outline the ideas behind our implementation of the synchronization approach. It is able to solve the linear program for data sets with $|J| \approx 100$, $|I| \approx 10^5$, and $|R| \approx 10^6$ on a standard PC within a few seconds.

Each row of A^T corresponding to a constraint (6.16) has exactly three non-zero entries and A is thus very sparse. The matrix A^T is closely related to the matrices arising in network optimization problems. In particular, it does not have full column rank. In the previous section the offset ambiguity has been introduced. Since we set $\bar{o}_1$ to zero, the corresponding column of A^T can be eliminated prior to the optimization. Our implementation checks for further redundancies that depend on the particular instance R and eliminates additional linearly dependent columns of A^T if existent.

We use a modern interior-point algorithm for our solver, a variant of Mehrotra's predictor corrector algorithm [Meh92] that is particularly well suited to handle the structure of (6.18). The primary advantage of interior-point algorithms versus the simplex method is that interior-point methods do not suffer from degeneracy of the problem. Practical implementations very rarely take more than 70 to 100 iterations to solve a linear program. In our case, the particular structure of (6.18) can be exploited making a single iteration very cheap. The concept of the algorithm as implemented here is based on Algorithm 14.3 in [NW99].

Apart from minor adjustments of the parameters proposed in [NW99] the main modification in our implementation concerns the storage format for the matrix A. Storing A directly would be extremely inefficient in terms of memory requirements as well as from a computational perspective. Our implementation comprises a specialized storage format for A, tailored to both the problem structure and the specific operations that appear in the interior-point algorithm. For matrices A arising from problem (6.18) this is a superior alternative to general purpose sparse matrix formats, as they are readily provided, e. g., by Matlab.

The main computational effort at each iteration of an interior-point algorithm is the computation and the Cholesky factorization of the matrix product $H = ADA^\mathsf{T}$. Here, D is a positive definite diagonal matrix that changes at each iteration. Due to our choice of setting up the variable y by first including $\widehat{T}$ and then $\bar{o}, \bar{r}$, the leading $|I| \times |I|$-block of H is a positive definite diagonal matrix, only the trailing $2 \cdot |J|$ rows and columns of H do have fill-in. This sparsity structure is also inherited by the Cholesky factor L of H. The leading $|I| \times |I|$-block of L can thus be computed in linear time. Given that typically $|I| \gg 2 \cdot |J|$, the computation of L and thus the solution of the overall problem is very cheap.

To demonstrate the huge gain in performance that is possible by using the tailored solver, we compare the runtime of our implementation with that of QSopt [ACDM] and SeDuMi [SRP]. QSopt is, as mentioned before, a solver that uses the simplex method. SeDuMi on the other hand is a Matlab interior-point code that, like our own solver, benefits from the special structure of H, but uses a more general—and therefore somewhat slower—storage format of the sparse matrix A^T, and a more general sparse Cholesky factorization.

Figure 6.1 shows the computation times for calculating the solutions of optimizations with 20 nodes and with 100 nodes, for different numbers of shared events. All measurements have been made on an AMD Athlon X2 BE-2300 CPU with 1900 MHz and 1 GB of main memory. From the figure it can be seen that our implementation actually works very well. The tailored solver brings a large performance gain—it reduces the computation time by typically at least a factor of 10–15.

Note that SeDuMi expects readily preprocessed input data in Matlab's sparse matrix format. The time needed for converting the data to this format is not included in the SeDuMi results in Figure 6.1. Especially for the larger problems, it can, however, be substantially higher than the time needed for solving the problem. The processing times shown for our own solver do include the time for reading the input data and preparing the optimization problem. For our specialized matrix format this step can be performed very efficiently; it accounts only for a negligible fraction of the total processing time.

In particular the results with QSopt underline that an off-the-shelf simplex solver is in fact highly unsuitable for the specific type of linear optimization problem that we deal with. Not only does the computation time grow rapidly with an increasing problem size, but also do the memory requirements. In contrast to that, our tailored implementation with its specific sparse matrix format is very memory efficient, and its observed runtime increases approximately linearly with the number of events $|I|$.

6.5 Properties of the MLE

Now that we have seen that it is in fact possible to calculate a solution of the linear program and thus the maximum likelihood estimator within reasonable time, we are interested in the quality of this solution. In this section we will thus tackle the question how good the synchronization result is.

With an increasing number of network packets that have been received by multiple network nodes the available amount of data to estimate the clock deviations increases. Thus, intuitively, one could expect that the quality of the estimate improves with the availability of more experimental data. Similarly, it sounds reasonable that it is very unlikely that the result of the time synchronization process is grossly wrong if the input data is very accurate. In this section, we confine ourselves to a simplified variant of

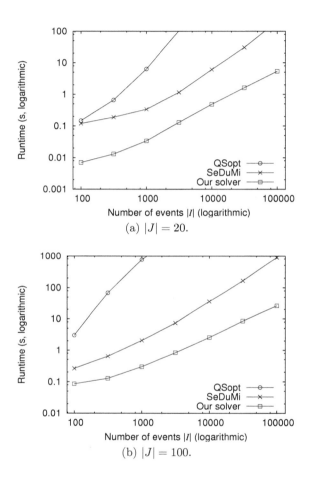

(a) $|J| = 20$.

(b) $|J| = 100$.

Figure 6.1: Performance comparison of QSopt, SeDuMi, and our own implementation as time synchronization LP solvers.

our estimator. For this simplified variant we can prove that these intuitive expectations are actually true. Since the complete proofs of these properties are quite complex and technical despite this simplification, we have not included them here. Instead they can be found in Appendix B. Below we will discuss the results and their implications, and we give a rough sketch of the proofs' ideas. Our numerical results presented later underline that the results also hold for the fully featured estimator with clock rate estimates.

The simplified estimator does not take clock rate deviations into account, i. e., it assumes that for each node the clock rate r_j is (approximately) 1 and thus correct with respect to the "true clock". Under this assumption the recorded time for a node-event pair $(i, j) \in R$ becomes $T_i + d_{i,j} + o_j$. Thus, the simplified maximum likelihood estimator, in analogy to the fully featured version, is the solution to the following problem:

$$\text{maximize } L = \prod_{(i,j)\in R} \lambda e^{-\lambda\left(t_{i,j} - \widehat{o}_j - \widehat{T}_i\right)}$$
$$\text{subject to } \forall (i,j) \in R : \widehat{d}_{i,j} = t_{i,j} - \widehat{o}_j - \widehat{T}_i \geq 0. \tag{6.20}$$

The optimum is again independent of λ and the above is equivalent to

$$\text{minimize } k(L) = \sum_{(i,j)\in R} \left(t_{i,j} - \widehat{o}_j - \widehat{T}_i\right) = \sum_{(i,j)\in R} \widehat{d}_{i,j} \tag{6.21}$$

under the same constraints.

Here we will point out two desirable properties for this version of the estimator. First, we give tight error bounds on the estimation error that hold under the assumption of a bounded timestamping delay. In particular this means that the algorithm does not amplify errors. Furthermore, we show that the estimator is consistent, i. e., for increasing data set sizes the estimate converges (in probability) to the true values of the estimated features. It thus supports the intuition that the estimate improves for a larger amount of observed and logged events in the nodes.

6.5.1 Error Bounds

In order to be able to give a bound for the estimation error we need to make two additional assumptions. While the first one guarantees network connectivity, the second one establishes an upper bound on the timestamping delay. Note that an upper bound for the timestamping delay does not constrain the practical applicability of the results presented here: for any practical experiment there is a finite number of receptions, and thus also a maximum timestamping delay.

The existence of the offset ambiguity as introduced in Section 6.3 prohibits that the absolute event times and clock offsets are determined from the log files. This also holds for the simplified estimator considered here. From the offset ambiguity, it is easy to see that there is also no way to estimate the relative time between two separate partitions within the same experiment. If there are no anchor points between two sets of nodes, there will be an ambiguity of the offset between these partitions. Thus, in order to get a bounded maximum estimation error, we need to assume network connectivity in the sense of anchor points: the network nodes do not fall into disjoint partitions, between which no events are shared.

Under such an assumption we can prove that

$$\forall j_1, j_2 \in J : |(o_{j_1} - o_{j_2}) - (\widehat{o}_{j_1} - \widehat{o}_{j_2})| \leq (|J| - 1) \cdot D \qquad (6.22)$$

if $D \in \mathbb{R}^+$ is an upper bound for the delays, i.e.,

$$\forall (i, j) \in R : d_{i,j} \leq D. \qquad (6.23)$$

Note that the bound is on the difference between two estimation errors because of the offset ambiguity.

The basic idea of the proof is the following. Consider two nodes j_1 and j_2. Then it can be shown that there always exists a sequence of unique nodes $s_1, ..., s_n$, $2 \leq n \leq |J|$, $s_1 = j_1$, $s_n = j_2$, with a special property. In this sequence, for each pair of subsequent nodes s_q and s_{q+1}, there is an event observed by both s_q and s_{q+1}, for which the estimated timestamping delay in s_q is zero. This, together with the nonnegativity of the timestamping delays, allows for the construction of an upper bound for $(o_{j_1} - o_{j_2}) - (\widehat{o}_{j_1} - \widehat{o}_{j_2})$. Since j_1 and j_2 can be chosen arbitrarily, the same bound also holds with j_1 and j_2 interchanged. This yields a corresponding lower bound for j_1 and j_2 and thus constrains the absolute value in the way given above.

We are also able to show that under the mentioned assumptions the bound is the best possible, i.e., that no estimator can exist that achieves a smaller worst-case error. This proof is based on two explicitly constructed worst-case scenarios that result in identical log files. The point is that although the resulting local log files are identical for the two scenarios, the clock offsets differ so much that no estimate can be better than the worst-case bound given above in both cases.

From the bound on the clock offset estimation error it is then quite easy to come to a similar bound on the event time estimates:

$$\forall i_1, i_2 \in I : |(T_{i_1} - T_{i_2}) - (\widehat{T}_{i_1} - \widehat{T}_{i_2})| \leq |J| \cdot D. \tag{6.24}$$

No part of the proof exploits the exponential distribution of the delays. Thus, independent of the derivation of the estimator, it shows that if there is an upper bound for the timestamping delays the estimates are close to the real values, regardless of the distribution of the delays within $[0, D]$.

6.5.2 Consistency

Differing from the results presented so far we will now no longer assume an upper bound on the timestamping delays. Instead we exploit their assumed exponential distribution. Under these premises, consistency of the simplified estimator can be established, which means convergence in probability to the correct offset values for an increasing number of observed events:

$$\forall j \in J : \plim_{|I| \to \infty} \widehat{o}_j = o_j + x, \tag{6.25}$$

where $x \in \mathbb{R}$ again comes from the offset ambiguity.

Similar to the previous results it is clear that such a result cannot hold if the network is not connected. We show the consistency of the simplified MLE under an additional regularity condition, defined as follows. We say that this regularity condition is fulfilled if there exists an undirected, connected graph $G = (J, V)$ and some positive constant β such that

$$\forall \{j_1, j_2\} \in V : E\left[\left|\{i \in I | \{j_1, j_2\} \subseteq R_i\}\right|\right] \geq \beta \cdot |I|. \tag{6.26}$$

This precondition can be seen as a somewhat stronger variant of the connectivity assumption used above. It is stronger in the sense that it requires an (in expectancy) ever-growing number of independent connections between all parts of the network with an increasing total number of observed events.

In order to prove the consistency result we show that the probability of the likelihood function having its optimum in an arbitrarily small environment around the correct clock offset estimates is arbitrarily high for a sufficing number of observed events. The key idea is to introduce a per-event decomposition of the objective function $k(L)$. Certain properties of these event-wise objective function terms form the basis of our proof. We have seen before that $k(L)$ is simply the sum of the $\widehat{d}_{i,j}$ for all (i,j) in R. Then a decomposition of $k(L)$ into event-wise components f_i is trivial:

$$f_i := \sum_{j \in R_i} \widehat{d}_{i,j} \qquad k(L) = \sum_{i \in I} f_i. \qquad (6.27)$$

We then switch our point of view. We regard the f_i no longer as functions of the estimated latencies, but as functions of the estimation error. It is then quite straightforward to show that all the f_i are convex and that they are all Lipschitz continuous with a common Lipschitz constant. Furthermore, we show that the expectancy for each f_i—as a function of the estimation error—has a global minimum for the correct estimate, and we give a non-negative lower bound for the difference between this expectancy in case of a non-zero estimation error and the minimum value. All these results in conjunction with the law of large numbers can then be used to establish the consistency of the estimator: for a given $\delta > 0$ there is a number of events N such that for $|I| > N$ the probability that the estimation error is greater than δ becomes arbitrarily small.

From the consistency result for the clock offset estimate it is easy to obtain a result on the quality of the event time estimates in the same asymptotic setting. If the estimation error of the clock offsets is close to zero (neglecting the offset ambiguity), the remaining event time error for an event i is $\min_{j \in R_i} d_{i,j}$. This minimum of the independent, exponentially distributed $d_{i,j}$ is itself exponentially distributed with parameter $|R_i| \cdot \lambda$. In particular this means that the expected estimation error decreases with the number of nodes observing the same event.

6.6 Numerical Evaluation

While the previous section assessed the performance of the proposed time synchronization method analytically, we will now focus on numerical experiments with the algorithm. In particular, we will show that the asymptotic properties that have been proven for the simplified estimator hold also for the fully featured version with clock rate estimates. Moreover, it will become clear that the convergence is quick enough to yield accurate estimates even for small event counts. Finally, we will show that the algorithm is robust if the assumptions—in particular the exponential distribution of the timestamping delays and the negligibility of clock drift—do not hold.

6.6.1 Methodology

Although desirable, using log files from a real testbed for an evaluation that rigorously quantifies the numerical quality of the calculations and the convergence speed is not possible: for real hardware, the correct values for the rates, offsets, and event times cannot be determined—this is why we need post-experiment time synchronization in the first place. Therefore, we use a two-step simulation in which the correct values are known. In the first step, the network is simulated to obtain globally consistent event times and a receiver relation R. Then, subsequently, we simulate the timestamping of the events in each node. Random clock rates, offsets, and timestamping delays are used to transform the correct timestamps, yielding a set of per-node log files. Like after a real experiment, our algorithm is then given these log files as input. The quality of the solution can be determined by comparing the results to the correct times, rates, and offsets.

Since our focus here is on supplying the numerical algorithm evaluation with an event set I, event times T, and a receiver relation R, rather than a rigorous performance evaluation of some protocol, we constrain ourselves to a basic simulation scenario. We use the network simulator ns-2 [ns2a], which has been extended to support promiscuous mode-like packet tracing: if a data packet could be successfully decoded by a node's simulated wireless interface, the packet is timestamped and logged, regardless of whether the node was the intended destination or just able to overhear the transmission.

In our simulations, $|J| = 100$ nodes move on an area of 1200×1200 meters according to the random waypoint mobility model. AODV [PR99] is used as a multihop routing protocol. Five pairs of nodes communicate continuously over a simulation time of 10 minutes, performing FTP data transfers over TCP connections. The IEEE 802.11 MAC protocol is used at a fixed network bandwidth of 1 MBit/s. The radio range is set to 250 meters, the carrier sense radius to 550 meters.

For the generation of the local node log files, the clock offset and rate of every node were chosen randomly. The choice of the offset is not at all critical: our implementation actually exploits the offset ambiguity to achieve improved numerical stability and, as its first step, shifts all processed log files to start at time zero. Consequently, whichever offset is chosen for a node, the performed calculations and thus the accuracy of the estimates are virtually identical. In our experiments, we sample the simulated offsets from a normal distribution with mean zero and standard deviation five seconds. For the

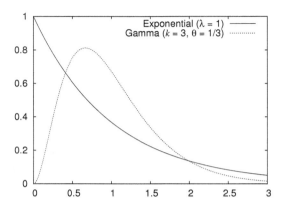

Figure 6.2: Probability density functions of exponential distribution ($\lambda = 1$) and gamma distribution ($k = 3$, $\theta = 1/3$).

clock rates, we used a gamma distribution[2] with mean one and different standard deviations. The gamma distribution has the advantages of yielding only positive rate values, and being concentrated around the expectancy. The probability density function of a gamma distribution is depicted in Figure 6.2. Unless otherwise stated, the parameters have been chosen to yield a standard deviation of 100 parts per million (ppm), which is rather pessimistic, meaning that on average a clock is wrong by more than eight seconds per day.

To be able to compare the synchronization results directly with the correct values from the simulation trace file, the rate and offset ambiguities need to be overcome. As stated in Section 6.3, the normalization constraints lead to scaled and shifted results if the average inverse clock rate is different from 1, and if the offset of node 1 is different from 0. This scaling and shifting can easily be removed by a linear-affine transformation after the optimization, based on the simulated average inverse clock rate and the offset chosen for the first node.

In the simulation, the arrival times of the same packet at different nodes actually differ

---

[2]The gamma distribution is given by the probability density function

$$f(x; k, \theta) = x^{k-1} \frac{e^{-x/\theta}}{\theta^k \cdot \Gamma(k)} \quad \text{for } x > 0,$$

where Γ is the gamma function. The gamma distribution has two parameters, called the *shape parameter* k and the *scale parameter* θ. It has mean $k \cdot \theta$ and variance $k \cdot \theta^2$.

slightly, because ns-2 simulates the radio propagation delay. For calculating the event time errors, we compare the estimated event time to the average ns-2 reception time. The differences are in the order of 10^{-7} seconds, and therefore significantly below the other errors that we are dealing with here.

6.6.2 Convergence and Numerical Accuracy

In our first set of experiments, we simulated the timestamping delays according to our assumptions, i. e., exponentially distributed. We varied the expected timestamping delay λ^{-1} between 10^{-3} and 10^{-5} seconds, and increased the number of events used for the synchronization. One central result in the previous section was that—at least for the simplified estimator—we can expect the quality of the estimates to improve if an increasing number of events is available for synchronization. In a practical implementation, numerical effects of, e. g., a limited floating point precision may influence the results. The primary purpose of the following simulations is to verify that this property also holds for our implementation of the full estimator, and to give an idea of the convergence speed.

Figure 6.3 shows the resulting average event time error with 95-percentile error bars. For better readability of the chart, only the upper part of the error bars is shown. The x-axis denotes the number of events that have been used for the synchronization. These have been chosen randomly from all transmissions with more than one receiver. Note that at the left hand side of the chart, for 100 events, there is only one sent packet per node on average. The randomly chosen clock errors are quite significant. Still, the synchronization eliminates them to an extent that allows for accurate event time estimates. If more events are available for the synchronization, the estimates improve further quickly, and the average event time errors are one order of magnitude below the timestamping delays. The convergence is so quick that for 1000 available anchor point events and more, only tiny fluctuations are left.

While the accuracy of the event time estimates is limited by the remaining part of the timestamping delay, this is not the case for the rate and offset estimates. Figure 6.4 shows how the average rate error develops in the same setting. The quick convergence for increasing $|I|$ is evident. The corresponding results for the offset estimates show the same behavior. For very small delays with $\lambda^{-1} = 10^{-5}$ seconds and a high number of available anchor points it can be seen that the accuracy does no longer improve linearly; then, the implementation approaches its numerical accuracy limits.

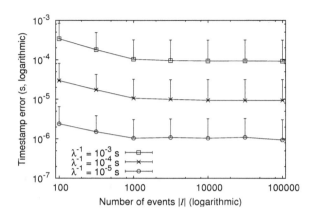

Figure 6.3: Event time errors depending on the number of events $|I|$ and the average timestamping delay λ^{-1}.

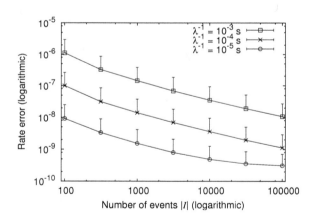

Figure 6.4: Rate errors depending on the number of events $|I|$ and the average timestamping delay λ^{-1}.

It also turns out that the estimation errors of rate and offset are largely independent from the true values of these parameters and their spread. For the offsets, this is relatively clear from the problem structure. For the rates, however, this trait is not immediately obvious. Tables 6.1 and 6.2 show the accuracy of the estimation results. The small deviations in estimation accuracy are remaining statistical fluctuations.

Another theoretical result from the previous section is that the event time estimation accuracy for an event i increases with $|R_i|$. More specifically, the result said that for correct rate and offset estimates, the remaining expected event time error is exponentially distributed with parameter $|R_i| \cdot \lambda$, and thus is $\lambda^{-1}/|R_i|$ on average. Figure 6.5 shows simulation results from log files with $\lambda^{-1} = 10^{-4}$ seconds and $|I| = 10\,000$. They exhibit exactly the predicted behavior. The chart shows the average event time error and again the 95-percentile upper error bar, where the events are broken down along the x-axis according to the number of nodes $|R_i|$ that observed them. The chart shows also the theoretical average error given by the function $x \mapsto \lambda^{-1}/x$, and it is evident that the results match the theoretical expectations very closely.

We may thus conclude that the convergence of the estimate is very quick, and a reasonable synchronization quality can be expected even if only a limited number of anchor points is available. The results also underline that the numerical performance of our implementation will not be the limiting factor for the accuracy in practical usage.

6.6.3 Robustness

So far, our simulations have used clocks and timestamping delays that match the assumptions made for the derivation of the approach. Now we assess how robust the estimator is if these assumptions do not hold. We thus use the very same estimator as before, but generate simulation data that intentionally contradicts the assumptions in different ways.

Table 6.1: Clock rate estimation error for different clock rate standard deviations ($\lambda^{-1} = 10^{-4}$ s, $|I| = 10\,000$).

std. dev. of rates	avg. rate error	95-perc. max rate error
10 ppm	0.0035202 ppm	0.0093509 ppm
100 ppm	0.0035785 ppm	0.0094587 ppm
1000 ppm	0.0035472 ppm	0.0092708 ppm

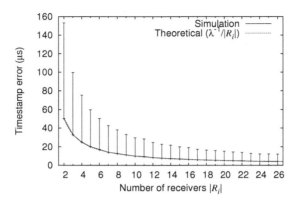

Figure 6.5: Theoretical and simulated event time estimation errors depending on the number of receivers $|R_i|$ ($\lambda^{-1} = 10^{-4}$ s, $|I| = 10\,000$).

The timestamp estimation errors occurring in these experiments are all shown in Figure 6.6. As a "baseline" for comparison it shows the results from the previous simulations, where all our assumptions hold, for $\lambda^{-1} = 10^{-4}$ s (labeled "exponential"). As could be expected, this achieves slightly better estimation results than all the cases where the assumptions do not hold.

First, we varied the distribution from which the timestamping delays were sampled. The result labeled "gamma" shows the estimation error for delays drawn from a gamma distribution with shape parameter $k = 3$. The scale parameter θ has been set to $1/(k \cdot \lambda)$. This yields a mean of λ^{-1} and therefore allows for a direct comparison to the results with exponentially distributed delays with the same mean. The probability density functions of these two distributions are shown in Figure 6.2, both adjusted to mean 1.

In the "multi-modal" simulations we assess the robustness to outliers in the timestamping delays. The majority of timestamping delays follows an exponential distribution

Table 6.2: Clock offset estimation error for different clock rate standard deviations ($\lambda^{-1} = 10^{-4}$ s, $|I| = 10\,000$).

std. dev. of rates	avg. offset error	95-perc. max offset error
10 ppm	$1.566\,\mu s$	$3.844\,\mu s$
100 ppm	$1.503\,\mu s$	$3.809\,\mu s$
1000 ppm	$1.500\,\mu s$	$3.958\,\mu s$

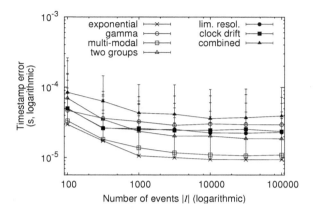

Figure 6.6: Event time estimation errors for increasing I if assumptions do not hold.

with $\lambda^{-1} = 10^{-4}$ seconds. 10 %, however, are instead drawn from a gamma distribution with $k = 10$ and a ten times higher mean. 1 % are "heavy outliers", sampled from a gamma distribution with a 50 times higher mean and $k = 100$. From these results, it can in particular be seen that our proposed method is very robust against outliers.

Inhomogeneous hardware with different timestamping delay characteristics is simulated in the "two groups" setup. The simulated network nodes are divided into two groups of 50 nodes each. The delays are exponentially distributed here, but the values of λ^{-1} differ by one order of magnitude: one half of the nodes uses $\lambda^{-1} = 10^{-4.5}$ seconds, the other half uses $\lambda^{-1} = 10^{-3.5}$ seconds.

We have also simulated the effects of a bad clock accuracy. As stated earlier, clocks in computer systems sometimes have a rather coarse resolution. In the simulations labeled "lim. resol.", the timestamping delays are again exponentially distributed with $\lambda^{-1} = 10^{-4}$ seconds, but the timestamps' resolution has been reduced to 0.1 milliseconds prior to performing the time synchronization. The estimator again deals very well with this effect. It is particularly remarkable that the availability of measurements from multiple nodes with different offsets allows for an estimation of the event times that is more accurate than the resolution of a single node's clock.

The "clock drift" simulations show the effect that randomly drifting clocks have on the accuracy of the estimates. Instead of linear clock functions, we use second order polynomials. Clock drifts are chosen independently from a Gaussian distribution with mean zero and standard deviation $3 \cdot 10^{-9}$. For our simulations, which cover a time span of ten minutes, this results in typical time stamp differences in the order of milliseconds to an otherwise equivalent clock with zero drift; the speed change of a typical clock sums up to several ppm during a ten-minute simulation. Nevertheless, as our results show, the effect on the synchronization accuracy is very limited.

Finally, the "combined" simulations incorporate all of the above sources of inaccuracies. In this data set, the timestamps are delayed according to the outlier-prone "multi-modal" distribution, there are two groups of nodes with different expected timestamping delays like in the "two groups" simulations, the simulated clocks drift as described above, and the timestamps' resolution is again limited to 0.1 ms. Even this combination of effects—all of which heavily contradict the foundations on which we have initially built our method—results in a degradation of the estimation quality by substantially less than one order of magnitude.

In summary, we conclude that the estimator is very robust and yields sensible results also if the various assumptions made for its derivation do not exactly hold. Although the quality of the estimates, as could be expected, degrades to a certain extent, they are still very good, and the estimator converges quickly to a high accuracy in all cases.

6.7 Real-World Experiments

The previously presented robustness assessment has shown that our proposed time synchronization method is able to deal well with a whole range of adversarial effects in the log data. Still, however, these evaluations were based on artificially generated simulation data. We will thus now complement them with an application of our method to real-world experimental data. While this, due to the unknown true values, does not allow to rigorously determine the remaining errors, it nevertheless provides a good intuitive understanding and shows how well the method can handle real data.

Our experimental setup consist of five PCs with rather heterogeneous hardware both in terms of CPU/memory and the wireless interface card. One of these nodes periodically broadcasts one packet per second, over a total of twenty minutes. The other four record and timestamp the received packets. Initially, the offsets have been reduced by approximately setting the clocks by hand.

In our figures, we use one of the receivers as a reference, and plot the differences in the recorded timestamps between this receiver and the other three. Figure 6.7 shows how— for unsynchronized clocks—this difference develops during the experiment. The almost exactly linear relative clock errors are clearly visible, as well as some timestamping delay outliers for nodes 1 and 2. The heterogeneity of the used hardware manifests itself in the fact that node 3 as well as the reference node do not produce that heavy outliers. (Exceptionally long timestamping delays in the reference node would result in parallel downward peaks.)

We have used our time synchronization algorithm on the data from the above experiment. This yields estimates $\widehat{r}_j$ and $\widehat{o}_j$ for the rates and offsets of the four receivers. We use those to eliminate the estimated linear clock deviations from Figure 6.7, by computing

$$\frac{t_{i,j} - \widehat{o}_j}{\widehat{r}_j} \approx T_i + d_{i,j} \tag{6.28}$$

for each timestamp. In Figure 6.8, we show the results of this correction. Again we plot the timestamp differences to the reference node, the y-axis uses the same scale as in Figure 6.7.

The approximation in (6.28) is exact if the estimates $\widehat{r}_j$ and $\widehat{o}_j$ of r_j and o_j are exact. Remaining clock deviations or estimation errors would therefore be visible in Figure 6.8: they would result in remaining timestamp differences to the reference node.

That such errors are in fact virtually non-existent becomes clear if we zoom the y-axis further in, as we do in Figure 6.9. It can be seen that the timestamping differences are typically in the order of some ten microseconds, with occasional outliers of up to 1–2 milliseconds. There is, however, no sign for a systematical (i.e., rate or offset estimation) error, like clocks drifting apart over time. This indicates that the rate and offset estimates are indeed correct.

Note that eliminating the estimated linear clock deviations according to (6.28) leaves the timestamping delays in the data. In a practical application, our approach would have been able to also eliminate long timestamping delays with very high probability. Recall that given exact rate and error estimates, it removes all but the shortest timestamping delay that occurred for events with multiple observers. Since we do not know the true times of the packet receptions in the experiment, we cannot tell how large exactly the then remaining deviations are. It seems, however, reasonable to assume that the smallest timestamping delay for a packet is within the same order of magnitude as the minimal timestamp difference to the reference node[3]. Considering (6.28), this difference is simply the difference of two timestamping delays. In the discussed experiment, the

---

[3]This does of course not hold when the minimum path delay is included in the timestamping delay. This, however, is not a problem here: recall from Section 6.2.3 that the minimum path delay in a node is equivalent to an additional clock offset, and may thus be eliminated.

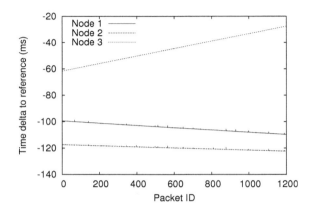

Figure 6.7: Unsynchronized timestamp differences in real-world experiments.

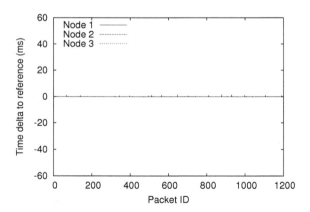

Figure 6.8: Synchronized timestamp differences in real-world experiments.

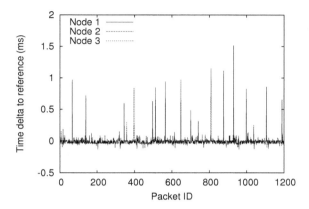

Figure 6.9: Synchronized timestamp differences in real-world experiments (zoomed).

average of the per-packet minimum timestamp difference is 19 microseconds, for 95 % of the packets it is below 49 microseconds.

6.8 An Alternative Approach Based on Least Squares

Instead of using a maximum likelihood estimator, it is also possible to perform anchor point based offline time synchronization based on least squares regression. This avoids the necessity to assume a specific distribution for the timestamping delays. We only outline this approach here, the details can be found in [JKM^{+}].

The basic idea of the least squares scheme is as follows. For an event $i \in I$ and $k, l \in R_i$ it follows from (6.2) that

$$
\begin{aligned}
r_l t_{i,k} &= r_k r_l (T_i + d_{i,k}) + r_l o_k \\
r_k t_{i,l} &= r_k r_l (T_i + d_{i,l}) + r_k o_l.
\end{aligned}
\tag{6.29}
$$

Subtracting the second from the first equation yields

$$
r_l t_{i,k} - r_k t_{i,l} = r_k r_l (d_{i,k} - d_{i,l}) + r_l o_k - r_k o_l.
\tag{6.30}
$$

In (6.30), the unknown value T_i is eliminated. Adding up these equations from all events shared by a pair k, l of nodes for which the set of shared anchor points $R^{k,l} := R^k \cap R^l$

is non-empty yields

$$\bar{t}_{l,k}\bar{r}_k - \bar{t}_{k,l}\bar{r}_l + |R^{k,l}|\bar{o}_l - |R^{k,l}|\bar{o}_k = \Delta d_{k,l}, \tag{6.31}$$

where

$$
\begin{aligned}
\bar{t}_{l,k} &:= \sum_{i \in R^{k,l}} t_{i,k} \\
\bar{t}_{k,l} &:= \sum_{i \in R^{k,l}} t_{i,l} \\
\bar{\Delta} d_{k,l} &:= \sum_{i \in R^{k,l}} (d_{i,k} - d_{i,l}).
\end{aligned}
\tag{6.32}
$$

For all pairs of nodes in the network that share common events, these equations are summarized in the matrix equation

$$M \begin{pmatrix} \bar{o} \\ \bar{r} \end{pmatrix} = \Delta d, \tag{6.33}$$

where $\bar{o}$ and $\bar{r}$ denote the vectors of the $\bar{o}_j$ and $\bar{r}_j$, respectively.

By attaching respective rows, normalization constraints to overcome rate and offset ambiguity just like the ones used above are added to the matrix. By a normal equations approach, the resulting equation system can then efficiently be solved in a least squares sense. This yields estimates for the rates and offsets of all nodes, from which, in turn, estimates for the event times can then easily be obtained. Because of the non-negativity of the delays, we may use the earliest resulting event timestamp after rate and offset correction have been applied:

$$\widehat{T}_i = \min_{j \in R_i} \frac{t_{i,j} - o_j}{r_j} = \min_{j \in R_i} (\bar{r}_j t_{i,j} - \bar{o}_j). \tag{6.34}$$

Convergence to the correct values for the rates and offsets—under appropriate connectivity assumptions similar to the ones used in Section 6.5—can be proven for the full least squares algorithm with rate estimates, with known error bounds. This is a significantly stronger result than what we have for the MLE approach.

It seems difficult to provide similar bounds analytically for the MLE. We may, however, numerically compare the results from the two estimators. For this purpose, we apply the same methodology as in Section 6.6. In Figures 6.10 and 6.11, rate estimation results for

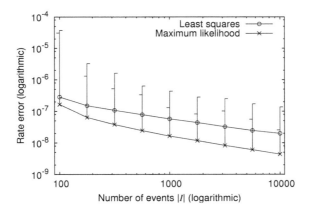

Figure 6.10: Rate estimation errors for exponentially distributed timestamping delays with least squares and MLE ($\lambda^{-1} = 10^{-4}$ s).

exponentially and gamma distributed timestamping delays are shown. The MLE yields better estimates for *both* these distributions. The differences are even more pronounced for the offset estimates. These are shown in Figures 6.12 and 6.13.

These results again underline that the MLE generally performs very well. It turns out that it produces more accurate estimates than the least squares approach even if the underlying assumptions are not satisfied. For the least squares approach, error bounds that do *not* depend on these assumptions can be proven. These theoretical bounds for the least squares estimator may thus be regarded as an indirect justification of the MLE.

6.9 Chapter Summary

In this chapter we have considered offline time synchronization for networks with local broadcast media. We have proposed a method to combine separate event log files from nodes in such a network into one single log file with a common time basis, in spite of deviating local clocks and latencies that occur when the timestamps for the events are generated. This is useful, for example, for the evaluation of experimental results. The key issue is how the deviations of the clocks and the latencies can be addressed without the necessity of additional communication between the network nodes.

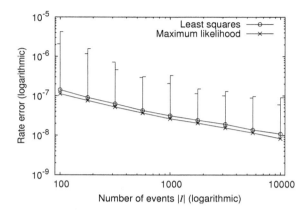

Figure 6.11: Rate estimation errors for gamma distributed timestamping delays with least squares and MLE ($\lambda^{-1} = 10^{-4}$ s).

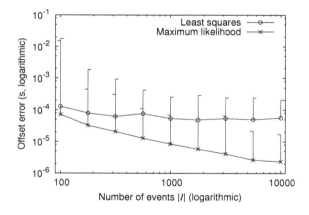

Figure 6.12: Offset estimation errors for exponentially distributed timestamping delays with least squares and MLE ($\lambda^{-1} = 10^{-4}$ s).

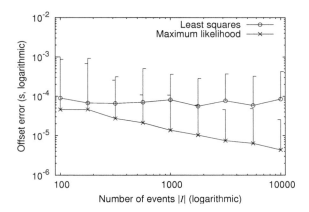

Figure 6.13: Offset estimation errors for gamma distributed timestamping delays with least squares and MLE ($\lambda^{-1} = 10^{-4}$ s).

Our algorithm utilizes transmissions that have been received by multiple nodes as anchor points. Such events allow for setting the clock readings of multiple clocks implicitly into relation, without exchanging any specifically time-related information. It has been shown how the synchronization can be formulated as an optimization problem, and how this problem can be expressed as a linear program. The special structure of this linear program can then be exploited by an efficient solution algorithm. We have presented an implementation of a specialized solver, and comparisons to other LP solvers underline the performance of the employed solution techniques. Furthermore, we have presented analytical results on the quality of the synchronization for a simplified variant of the estimator. In particular, these results include a bound on the maximum possible synchronization error, depending on the maximum timestamping delay, and a consistency proof. A subsequent numerical evaluation has shown that the convergence to accurate estimates is quick, and that the solver is robust even if the underlying assumptions on the distribution of the timestamping delays do not hold. The application of our method to real-world data and a numerical comparison to an alternative scheme that is based on weaker assumptions further underlined its performance.

We consider the presented approach generally applicable whenever event data is distributed over multiple sources, and common events can be used as anchor points for a maximum likelihood time synchronization. Apart from supporting the interpretation of experimental results in networks with local broadcast media that we have primarily considered, we envision adaptations of the proposed technique for example in the field

145

of network forensics, where data of, e. g., multiple intrusion detection systems (IDS) or firewall logs is combined. There, too, certain events will often have been observed in parallel by multiple systems.

Chapter 7

Conclusion

In this work, we have discussed the application of implicit feedback and coordination in wireless multihop networks. We have devised solutions to a number of challenging questions in such environments, all making use of information that has not explicitly been transferred or requested: exploiting the very properties of the local broadcast medium can turn difficulties into opportunities.

The first application of implicit feedback, presented in Chapter 2, was the implicit hop-by-hop congestion control principle, with its realization in the CXCC protocol. CXCC establishes hop-by-hop backpressure towards the source of a packet stream by tightly limiting the number of packets queued in each intermediate node. This is achieved by overhearing the forwarding of the next hop node to implicitly determine its queue state, and by a set of simple, localized packet forwarding rules. The protocol does not require to maintain any rates, window sizes, or similar parameters, neither at forwarders nor at the source. The RFA mechanism has been introduced as a light-weight mechanism to achieve single-hop reliability with very low overhead. It avoids any unnecessary packet payload retransmissions.

Using CXCC as a basis, we have subsequently extended the application of implicit feedback in three directions. The first one, end-to-end unicast reliability, was pursued in Chapter 3, in the BarRel transport protocol. It exploits a known limit on the number of packets in transit to infer successful end-to-end packet delivery. The combination of BarRel and CXCC can provide TCP-equivalent service without multihop feedback from the sink to the source. It obtains information on the end-to-end arrival of information implicitly, and hence refutes the widespread intuitive belief that end-to-end acknowledgments are indispensable for end-to-end communication with TCP-like reliability.

In the BMCC protocol, which has been discussed in Chapter 4, implicit hop-by-hop congestion control has been generalized to the multicast communication pattern. BMCC has been implemented in combination with the geographic multicast routing protocol SPBM. Because SPBM provides only heavily aggregated information on the receivers in both source and forwarders, congestion control is particularly difficult to perform in such a context. The two central challenges were the generalization of the RFA/(N)ACK handshake to multiple local receivers on the one hand, and the relaxation of the backpressure mechanism to overcome the crying baby problem on the other hand. The backpressure pruning mechanism was introduced to solve the latter issue. It combines overheard information on packet forwarding with link status feedback in KAL packets to identify branches of the multicast tree with heavy load indirectly, then building up backpressure in a selective manner.

The last application of implicit feedback and co-ordination to a network protocol was noCoCo, introduced in Chapter 5. noCoCo is a scheme for the local co-ordination of the packet transmission order which maximizes the number of opportunities for network coding within bidirectional data flows. It goes beyond pre-existing network coding schemes for the wireless domain in that it jointly encompasses coding and scheduling decisions. It therefore does not need to rely on coding opportunities arising spontaneously. Nevertheless, it also does not need any explicit co-ordination between endpoints or intermediate nodes. Instead, it is based on overhearing transmissions performed by neighboring nodes, and on the basic packet forwarding rules that had already proven their suitability in CXCC and BMCC. In fact, noCoCo can be seen as a coding-aware extension of CXCC.

The last contribution of this thesis, the post-facto offline time synchronization algorithm presented in Chapter 6, exploits information that can be obtained from a set of log files resulting from experiments performed with nodes in a network with local broadcast characteristics. The parallel reception of the same transmission by multiple receivers yields information on the relation of their clock readings at the time of reception. Such "anchor points" can be used to overcome the problem of deviating clocks in the evaluation of experimental results. In that chapter, implicit feedback was thus not used in the design of a protocol. Instead, implicitly obtained information formed a basis for the inference of unknown parameters that can neither be directly observed, nor measured online in the network without necessitating additional traffic.

In summary, each chapter considered the application of knowledge about the specific properties of a wireless multihop medium. This knowledge has been used to derive information, the explicit transmission of which would otherwise have been costly. Some of the addressed issues were previously known, and there were pre-existing solution approaches, others have been pointed out and tackled for the first time. The resulting approaches have demonstrated that making use of the information that is available anyway can lead to novel and sometimes unconventional, yet highly effective solutions.

We are convinced that the concept of implicit feedback has potential far beyond the applications shown here. Hence, we hope that our discussion will motivate other researchers to recognize the possibilities offered, and not only the difficulties caused by a given environment—open your protocols' eyes and let them "read between the packets".

Appendix

Appendix A

Detailed Overview of MANET Transport Protocols

Many unicast transport protocols have been proposed for wireless multihop networks. Only those sharing some similarities with the congestion control and reliability protocols introduced here have been discussed in Chapters 2 and 3. In this appendix, which has in similar form also been published as a journal article [LSM07b], we give a broader overview.

We introduce a structure on the proposed solutions, grouping them in categories based on the key problem they focus on. Section A.1 describes approaches that deal primarily with problems due to mobility-induced route failures. Random wireless losses are the main concern of the proposals in Section A.2. In Section A.3 approaches are discussed that deal with the special properties of a shared medium. Self contention between data- and ACK-traffic is the main concern of the approaches in Section A.4. In Section A.5 we describe proposals that artificially limit the packet output of TCP in order to adapt TCP to MANETs. Finally, there are a number of approaches specifically tailored to the properties of MANETs. These are not based on TCP. They are discussed in Section A.6. Some approaches can be assigned to multiple sections, since they address more than one key problem. In these cases we present them in the section where they provide the most significant contribution.

A.1 Dealing with Route Failures

In a typical mobile ad-hoc network route failures occur frequently. The amount of time required for discovering a new route has a negative impact on the standard TCP congestion control mechanism. For a long time no packets are delivered and no acknowledgments are received, causing the TCP sender to reduce its window size dramatically, even though in fact no real congestion situation might exist. The approaches described in this section primarily tackle this problem.

One of the very first approaches to deal with the congestion control problem in MANETs at all is called **TCP-Feedback (TCP-F)**. It has been proposed by Chandran et al. in 1998 [CRVP98], and has become the basis for a lot of subsequent work in this area.

The authors of TCP-F propose to disable the TCP congestion control mechanisms in case of network-induced, non-congestion related losses or timeout events caused by route failures. The TCP sender is notified explicitly when routing failures occur (RFN, Route Failure Notification) and when a new route has been discovered (RRN, Route Re-establishment Notification). When a route to the destination node is currently not available, i.e., after receiving an RFN message, the TCP-F sender enters a "snooze" state. In this state it freezes its TCP state values such as timers and window sizes.

An intermediate node generates an RFN message when it detects a link failure on the route. Once a node previously generating or forwarding an RFN learns about a new route to the destination node, it generates an RRN message and sends it to the source node. When the source node receives an RRN, it resumes the TCP session with the previously frozen state values. To prevent a TCP-F session from remaining in snooze state indefinitely in case an RRN message gets lost, an additional timeout is used as a fallback.

Holland and Vaidya analyze the impact of **Explicit Link Failure Notification (ELFN)** techniques on TCP performance [HV99]. ELFN means feedback from lower layers to notify TCP explicitly about link or routing failures. In case of such a failure, the sender enters *standby mode*, which is the equivalent to TCP-F's snooze state.

In contrast to the TCP-F proposal no explicit notification in case of a re-established route is used. Instead, a TCP-ELFN sender sends probe packets in regular intervals when in standby mode. A probe packet is not a special control packet. Instead the first data packet in the send queue is used for that purpose. Standby mode is left as soon as a probe packet is acknowledged by the destination node.

Likewise, for route failure notifications no special control packet is introduced in ELFN. The authors propose to either piggyback the notification message onto a route failure message sent by the routing protocol (as used, for example, in DSR), or to use an ICMP host unreachable message for that purpose.

ELFN has become very well-known and has served as a basis for many later approaches.

Monks at al. showed in [MSB00] that although ELFN can indeed improve the performance of TCP, there are situations where a severe performance degradation is possible. This is particularly true for scenarios with many active connections. The reason is that an ELFN-like mechanism makes TCP behave more aggressively, i. e., a higher number of packets is present in the network. Therefore the MAC layer contention is generally higher. This in turn leads to more collisions, a higher packet loss rate on the MAC layer and eventually to false broken link detections; therefore wrong route failure notifications are sent and unnecessary route discoveries are performed. This observation is quite fundamental and should generally be taken into account in mobile ad-hoc networks. It does not only apply to ELFN or TCP-F, but also to many of the following approaches.

In **TCP with BUffering capability and Sequence information (TCP-BuS)** by Kim, Toh and Choi [KTC00], the idea of explicit notifications from the routing layer is extended. Additional measures are proposed to improve TCP performance.

In TCP-BuS, intermediate nodes buffer packets in case of a route failure instead of dropping them; the intention is to not having to retransmit all these packets. To avoid timeouts at the source when buffered packets are delivered after the route is reconstructed, the timeouts for these packets are extended. Selective retransmission requests by the destination node allow a fast recovery from lost buffered packets without the need to wait for these extended timeouts to expire. A scheme for avoiding unnecessary retransmission requests is proposed.

In order to ensure the delivery of control messages like route failure notifications, a reliable delivery scheme is suggested for these messages. To detect a re-established route both explicit notifications and probe packets, i. e., both the TCP-F and the ELFN way of detecting the new route, are examined.

In [DB01] Dyer and Boppana compare the performance of different TCP variants over three MANET routing protocols. Based on their finding that an exponentially growing retransmission timeout (RTO) is problematic in MANETs because of the route failures, they propose to keep the RTO fixed after the first retransmission. Their **Fixed RTO** scheme thus leads to a periodic sending of packets. Essentially this has the same effect as ELFN's probe packets. However, their scheme is much simpler because it does not interact with the routing layer, and due to the dependency of the RTO on the measured round trip time it is inherently adaptive

They compare this mechanism with the use of TCP's selective acknowledgment (SACK) and delayed acknowledgment (DACK) extensions. The results of their ns-2 simulations show that the delayed and selective acknowledgment mechanisms of TCP yield only minor improvements, but with fixed RTO a performance increase similar to that of ELFN is possible.

In the **ENhanced Inter-layer Communication and control (ENIC)** scheme proposed by Sun and Man [SM01], too, ELFN-like route failure handling and the TCP SACK and DACK mechanisms are combined. In comparison to TCP-BuS, ENIC requires less assistance from the intermediate nodes. Like in ELFN, no separate notification messages in case of a route failure or a recovered route are transmitted. Instead the reuse of notifications inherent to the routing protocol is proposed. A buffering of packets at the intermediate nodes is not suggested, queued packets are dropped when a route breaks.

Not only the sender, but also the receiver is notified of route failures in ENIC. Like the sender it freezes its state, particularly that of the DACK mechanism. With a broken route, acknowledgment packets would not be able to reach the source node, anyway.

In ENIC, a new retransmission timeout value (*Temporary RTO*) is calculated after a route change in a heuristic fashion. It is based on the hop counts of the old and the new route. Thus of the approaches described here ENIC is the first one taking potential changes of route characteristics after a route reestablishment into account.

Zhou et al. in [ZSZ03] focus especially on the problem of changing route properties after a route reestablishment. As an extension to the ELFN mechanism they propose to recalculate TCP's congestion window size (cwnd) and slow start threshold (ssthresh) parameters based on the properties of the new route. They call their approach **TCP-ReComputation (TCP-RC)**.

In TCP-RC, the route length and the round trip time—both relative to their previous values—are used to determine new values for cwnd and ssthresh. In ns-2 simulations they compare TCP-RC to plain TCP Reno and find performance improvements; however, it is not clear how much of this improvement is due to the ELFN mechanism alone, since this variant is not examined.

With **TCP for mobile ad hoc networks (ATCP)** Liu et al. [LS01] present a solution to several TCP problems in MANETs. ATCP not only handles route failures, but it is also intended to correctly handle longer periods of disconnection and to distinguish congestion-related from other losses. For the latter explicit congestion notification (ECN) messages are used.

Instead of invoking standard TCP congestion control in the case of non-congestion losses ATCP uses a TCP state freezing mechanism with an TCP-ELFN-like probe packet mechanism. A loss is considered to be non-congestion related if no ECN message is received.

An important point for the authors is to preserve standard TCP compatibility as far as possible. Therefore ATCP is implemented in an additional layer below the transport layer, reducing the interaction with TCP to a minimum.

A problem with explicit link failure notification schemes observed by Yu in [Yu04] is that still a number of data packets and ACKs may get lost before the state is frozen. This has negative effects *after* the state is restored: missing packets or missing ACKs will then cause timeouts or duplicate acknowledgments. Yu's **cross-layer information awareness** scheme overcomes this by extensively using cached route information from the DSR routing protocol. Two mechanisms are introduced, *Early Packet Loss Notification (EPLN)* and *Best-Effort ACK Delivery (BEAD)*.

EPLN notifies TCP senders of the sequence numbers of lost packets that could not be salvaged. The sender can then disable the retransmission timer and retransmit the respective packets upon route reestablishment. BEAD generates ACK loss notifications at intermediate nodes and sends them towards the TCP receiver. This prevents ACKs from being permanently lost. A node forwarding such a loss notification may send an ACK with the highest affected sequence number to the TCP sender if it is able to do so. Otherwise the TCP receiver will retransmit an ACK with the highest sequence number when a new route is present.

Wang and Zhang—in contrast to most of the previously described approaches—propose a pure end-to-end mechanism. It is called **TCP with Detection of Out-of-Order and Response (TCP-DOOR)** [WZ02]. They regard the deployment and maintenance of cross-layer solutions as too intricate in many scenarios.

In TCP-DOOR, data packets and/or ACKs delivered out-of-order are utilized as an indication of changed routing without the need of explicit feedback. While the receiving node can notify the sender about detected out-of-order data packets, the sender itself may notice ACKs arriving out-of-order.

Two mechanisms are suggested as a reaction to the occurrence of out-of-order events. When out-of-order packets are detected, the sender may temporarily disable TCP's congestion control mechanisms by keeping its state variables constant. Additionally, it may fall back to a somewhat older state. Thereby it reverts effects of congestion control mechanisms that might already have occurred. This last mechanism is called *Instant Recovery*. The expected effect is similar to that of freezing: after the reset "wrong" changes to TCP's parameters are reverted, and the connection continues to operate as if no route change had occurred.

Through simulation studies the authors have found that it is perfectly sufficient to detect out-of-order events either at the sender or at the receiver. A combination of both detection mechanisms does not yield significantly better results. The best performance is achieved by a combination of both possible reactions, that is, by temporarily disabling congestion control and performing Instant Recovery.

The authors recommend their approach primarily for mixed ad-hoc and wired network scenarios, where it would be particularly hard to adopt a feedback-based approach. Where possible, they suggest to use some feedback-supported method instead.

In [GAGPK03] Goff et al. propose an early detection scheme for route failures. The idea of **preemptive routing** is to foresee route failures before they actually occur, and to initiate a new path discovery early. This is intended to avoid or at least to reduce disconnection times. Especially for TCP—with the negative effects that the disconnection periods have on its performance—significant improvements are expected by the authors.

Upon reception of a packet each node along a route looks at the received signal strength. If it is below a given threshold, a warning is sent to the source node of the route. To mitigate short-term effects like small-scale fading, the use of an exponential average or of a verification of the measurement by sending some small ping-pong packets along the respective link are proposed. In simulations they evaluate their scheme and find a largely improved TCP performance when preemptive routing is used.

In [APSS04] Anantharaman et al. perform an analysis of TCP behavior in mobile ad-hoc networks, studying the factors that influence the protocol's behavior. They conclude that the schemes proposed so far, in particular the ELFN scheme, are only able to deal with part of the problems TCP is facing due to breaking links and route failures. They propose a set of three mechanisms with a somewhat broader scope, designed specifically for the DSR routing protocol [JM96]. They call the combination of these mechanisms the **Atra framework**.

The main aims of Atra are the minimization of route failures, their prediction and a fast notification of the source in case of a route failure. The mechanisms used to achieve these goals are called *Symmetric Route Pinning, Route Failure Prediction,* and *Proactive Route Errors.* Symmetric Route Pinning forces TCP acknowledgment packets to use the same route as the corresponding data packets. Usually, different routes could be used in DSR. The authors' reasoning is that using different routes increases the probability of a route failure—there are two routes that may fail.

The Route Failure Prediction mechanism in Atra works like the preemptive routing scheme: each node along the route estimates the trend of received signal strength values. In case the Route Failure Prediction doesn't work, the third mechanism, Proactive Route Failures, notifies the sources of *all* connections that use the broken link. This is different from standard ELFN mechanisms, where only the source of the packet that could not be transmitted on the MAC layer is notified of the problem, meaning that every connection sharing the broken link has to detect the problem separately.

Another approach anticipating route failures is the **signal strength based link management** by Klemm et al. [KYKT05]. They point out that the 802.11 MAC cannot identify link breaks correctly in case of congestion. The situations when two nodes move out of transmission range and when a congested area does not allow a successful RTS transmission cannot immediately be distinguished. The authors aim to distinguish these cases better and to provide appropriate reactions in conjunction with the AODV routing protocol [PBRD03].

The first key idea is the same as that of preemptive routing: a history of node distances, estimated by received signal strength information (RSSI), is kept for the neighbors. If a used link is about to break the routing protocol is informed and a new route to the destination can be searched early. In addition to the signal strength based link management it adds mechanisms to ease the salvaging by temporarily increasing the transmission power. In case congestion prevents the transmission of data, the number of RTS/CTS retries before a is packet dropped is increased.

The main difference between route failure anticipation schemes and the approaches discussed before is that these schemes try to avoid performance degrading effects from occurring in the first place, instead of alleviating their effect. This concept is also the key idea in the **multipath TCP** scheme by Lim et al. [LXG03]. They use existing multipath routing protocols to study the effects that these approaches have on the TCP performance. They compare two modes how the additional routes provided by the routing layer can be used: distributing the TCP packets to multiple routes in parallel and maintaining additional routes just as a backup in case of a route failure.

By simulation with Fixed RTO TCP (see above) as a transport protocol they find that it is not beneficial to use multiple paths in parallel. Using different paths—with different round trip times—makes TCP's RTT measurements unreliable. Additionally, out-of-order delivery effects produce many duplicate ACKs and thus trigger unnecessary congestion control reactions.

In contrast to that, maintaining one additional route as a fallback has a positive effect. In case of a route error the transmission can be continued quickly. The authors call this scheme *backup path multipath routing*. In some sense, it can be seen as a consequent variant of the route failure prediction schemes: instead of establishing an alternative path when a route failure is about to occur, it is maintained from the beginning, in order to be always quickly available.

A.2 Coping with Wireless Losses

A wireless link per se is much more prone to more or less random packet losses than a wire-line connection. Such losses are detrimental for a transport protocol's performance if they are misinterpreted as congestion-induced packet drops.

This problem is not a central issue when 802.11-like MAC protocols are used, where the link layer provides single-hop reliability. In this case packets are dropped by the link layer only after a number of failed transmission attempts or failed RTS/CTS handshakes. This in turn usually happens when either a link is lost or when a lot of packet collisions occur. The latter means that there probably is a lot of traffic in the area around the currently forwarding node and thus congestion. Therefore although the packet loss has not been caused by a queue overflow in a router it can still be a valid congestion indicator.

However, it can well be argued that there might be networks without single hop reliability. In such a case the proposals presented in this section show how end-to-end transport protocols—designed to rely on missing packets as congestion indicators—can learn to deal with random losses. Some work in the area of distinguishing wireless losses from congestion losses has also been done in the area of wired-cum-wireless networks, see for example [BV05]. Here, we only focus on the approaches for mobile ad-hoc networks.

Güneş and Vlahovic propose to introduce three states in TCP senders [GV02]. This approach is called **TCP with Restricted Congestion Window Enlargement (TCP/RCWE)**. It is based on the Explicit Link Failure Notification mechanism.

In TCP/RCWE, link breaks and thus the corresponding losses are handled by ELFN. In addition, RCWE aims to deal well with random losses. To this end, the authors propose a mechanism based on a heuristic observing the value of the Retransmission Timeout (RTO). If the RTO increases the congestion window size is not increased. If the RTO decreases or remains constant the congestion window size is increased according to TCP's rules.

In ns-2 simulations, RCWE is found to cause a much smaller congestion window, leading to higher goodput and less packet losses. But it is compared only to standard TCP without ELFN, so it is not clear how much of the performance gain is due to ELFN alone. Since seemingly IEEE 802.11 with link layer reliability is used in the simulations, the observed performance gain might also occur only because cwnd is increased less often and thus is smaller on average. In this case, the observed effect would be the same as the one discovered later by Fu et al. in [FZL+03], leading to the congestion window size limitation schemes described in Section A.5.

To improve the reliability of bit error loss detection, Fu et al. in [FGML02] combine multiple metrics instead of relying just on a single one. They call their approach **ADTCP**. They state that the main problem of end-to-end transport protocols in mobile ad-hoc networks is the noisiness of the measurements of indicators for certain network events.

Two metrics are proposed to detect network congestion. The *inter-packet delay difference* at the receiver, defined as the time elapsing between two successive packet arrivals, increases when congestion occurs. Additionally, the *short-term throughput*, describing the throughput in a certain time interval in the immediate past, decreases in case of congestion. These two metrics are combined to gain a more robust congestion indicator. In a similar way, out-of-order packet arrivals and the packet loss ratio are used to detect route changes and channel errors.

In ADTCP, the receiver detects the most probable current network state and includes this information into its feedback to the sender. Both sender and receiver behavior are altered appropriately, but at the same time remain compatible with standard TCP.

An ADTCP sender can talk to a standard TCP receiver and vice versa. ADTCP also behaves TCP-friendly.

Especially because of these last properties ADTCP might be an interesting option for scenarios with different protocols in the same MANET, with mixed wired and wireless infrastructure, or with existing TCP-based applications.

For mobile multimedia communication Fu et al. also propose an adaption of TFRC called **ADTFRC** [FML03]. TFRC is a rate-based transport protocol originally developed as a TCP-friendly transport protocol for wired networks with smooth rate adaption properties [FHPW00].

ADTFRC applies the same ideas to TFRC that ADTCP applies to TCP. An identical combination of metrics and general mechanism are used to distinguish loss types and to provide receiver-based feedback. ADTFRC shares most of the benefits of ADTCP, especially the possibility of incremental deployment—ADTFRC, too, is compatible to standard TFRC communication partners.

De Oliveira et al. propose to use the measured round trip times to distinguish between congestion and medium losses [dOBH03]. In their **edge-based approach** the TCP congestion control reaction is circumvented if a medium loss is detected. Also, route failures are detected when a timeout occurs and no packets at all have been received for a longer period of time. In this case TCP enters an ELFN-like "probe mode", in which packets are transmitted at regular intervals in order to be able to detect a re-established route.

The specific mechanism to distinguish congestion from medium related losses is presented in [dOB04]. It is based on fuzzy logic [Zad65, Zad68]. The authors use ns-2 simulations with a single observed flow and some background traffic to examine the performance of their approach, and find a good detection accuracy, although the amount of samples needed can be high, causing the algorithm to be too slow for highly dynamic scenarios. The authors name some techniques that could be used to improve the fuzzy logic engine in this regard.

Given the approaches described in this section, the results presented in [CXN03, FZL$^+$03] are of particular interest. The authors show that the assumption that more knowledge on the cause of packet losses solves the problems of TCP in MANETs may not be true. On the contrary, the resulting, more aggressive TCP behavior might lead to an even higher load on the network and thus more congestion problems.

A.3 Managing a Shared Medium

In a wireless network the medium is shared by all nodes in a certain area. Dealing with this property is a big challenge when one wants to perform congestion control in such networks: it makes congestion a spatial phenomenon, happening no longer in a node, but rather in an area. Several mechanisms paying attention to these special limitations have been proposed. They are presented here.

In a frequently cited work by Fu et al. [FZL+03] the authors show that for a given topology and traffic pattern there exists an optimal TCP window size, but TCP is unable to find it. Instead it uses larger windows, leading to dropped packets caused by link-layer contention. This observation has influenced research in wireless multihop congestion control significantly. Given this background the authors propose two mechanisms to improve TCP by earlier reaction to link overload—a distributed **Link RED (LRED)** and an **adaptive pacing** strategy.

The Random Early Detect (RED) mechanism in wired networks [FJ93] drops packets in router queues at random with a probability that increases linearly with the queue length. This makes TCP flows passing through the router regulate their bandwidth requirements before severe congestion occurs. LRED's intention is to make the TCP flows regulate their window size closer to the optimal region for MANETs. In LRED, the probability for a drop is based on the observation of the number of transmission attempts needed on the MAC layer. The mechanism is enabled once a certain threshold is exceeded. With more frequent retries the probability to drop a packet is increased, because this situation is interpreted as a sign for local congestion.

The other technique—adaptive pacing—is also enabled once the LRED retransmission count threshold is reached. When active, the mechanism adds an additional packet transmission time to the sender's MAC backoff timer. Therefore the medium can be expected to be free also at the next node downstream, avoiding the so-called *exposed receiver problem*. This leads to a two hop co-ordination mechanism as the sender waits long enough that a packet can be forwarded from the receiver one hop further.

Xu et al. in [XGQS05] focus especially on TCP unfairness problems caused by the locally shared medium. Their proposal to improve TCP fairness, **Neighborhood RED (NRED)**, is like the previously discussed Link RED based on Random Early Detection.

In NRED, every node estimates the number of packets queued in its neighborhood in total, i.e., in all nodes in the vicinity. All these packets form a virtual, distributed neighborhood queue. If the length of this queue exceeds a threshold, packets start getting dropped with increasing probability.

NRED consists of three steps. The neighborhood queue size estimation is performed by analyzing the channel utilization. Therefore the transmissions of the neighbors are overheard. If the utilization exceeds a threshold the neighborhood is presumed to be in an early congested state. A drop probability is calculated and is sent explicitly to all neighbors to inform them. Each node calculates its own drop probability based on the received notifications. Incoming packets are dropped with this local probability, in total leading to RED-like packet dropping in the virtual neighborhood queue.

The authors of **COntention-based PAth Selection (COPAS)**, de Cordeiro et al. [dMCDA02], focus on a problem of TCP in MANETs called the *capture problem*. Nodes can capture the medium unfairly and gain an advantage in comparison to others.

COPAS is an extension for reactive routing protocols. During route discovery all routes between a source and destination node are gathered. Then two disjoint routes are used to forward upstream TCP traffic and downstream acknowledgments respectively, in order to avoid effects where one of the two directions captures the medium. The decision which routes are chosen is based on congestion measurements performed during the discovery process. The measurements are based upon the backoff times for which the node had to wait before the medium became free. They are updated continuously during operation, and a route that becomes too congested is substituted by a better one.

Interestingly, the fact that two disjoint routes are used for forward and backward traffic is perfectly opposed to the Symmetric Route Pinning technique from the Atra Framework (see Section A.1), where special care is taken to use the *same* route.

Ye et al. in [YKT04] propose to extend the routing in order to separate flows spatially on the basis of distributed congestion information in their **Congestion Aware Routing (CAR)** approach. In contrast to most previous work in the congestion-aware routing area they focus on TCP flows and the interaction with the congestion control mechanism.

First, they evaluate the theoretical benefit for spatial separation by simulating a centralized approach (Centralized CAR, CCAR). They assume that every node has total knowledge about source, destination, and route of every single TCP flow.

164

A decentralized approach (Distributed CAR, DCAR) is then described as a more realistic scenario. Every node locally calculates a congestion weight representing its local load situation and broadcasts it to its neighbors. The AODV routing protocol is used in an adapted form. Route discovery messages are collected for some time until they are forwarded, and routes are chosen based on the path's aggregated weight. The destination node uses the path with the minimal weight to send the route reply message.

The performance analysis shows that both approaches outperform shortest path routing protocols for long paths in terms of throughput. The centralized approach clearly outperforms the distributed mechanism and is also favorable for short paths where the distributed approach fails. However, for obvious reasons it is not feasible in practice. As the main reason for the inferior performance of DCAR in comparison to CCAR the overhead for broadcasting the—potentially already outdated—congestion information has been identified by the authors.

Spatial separation of flows as in COPAS or CAR might actually be an effective means of reducing contention-related negative effects in MANETs. However, problems could arise in such an environment due to possible interactions between congestion control and routing. In particular feedback-loop oscillations can occur if congestion induces route changes, thus create network load in different areas of the network, and eventually force other streams to change their routes.

A completely different approach to solve TCP's problems in MANETs has been proposed by Kopparty et al. in [KKFT02]. In **Split TCP**, congestion control and end-to-end reliability mechanisms are separated.

In intermediate nodes on the path between sender and receiver so-called *TCP proxies* are automatically established. The proxies subdivide the path into several independent segments. Each proxy buffers packets and transmits them either to the next proxy or to the final destination. *Local acknowledgments* are used to acknowledge packets within one segment. In addition to local acknowledgments end-to-end ACKs—potentially cumulative—are used to ensure reliability in case of a proxy failure.

Split TCP intends to alleviate mobility related effects by keeping other path segments functioning if a link breaks in one segment. Capture effects are expected to be less severe, since the transportation of data occurs in shorter stages and thus the size of the captured region is reduced.

Unlike most previously described approaches, Split TCP does not try to modify TCP behavior to introduce "knowledge" about specific properties of MANETs into the mechanisms. Instead, the long way for feedback from the sink back to the source is identified as a major problem and alleviated by shortening the feedback path.

The focus of Berger et al.'s scheme for **alleviating self-contention** [BYS$^+$04] is on contention between packets belonging to the same transport layer flow. This can happen for both packets going into the same and packets traveling in the opposite direction.

The authors propose two mechanisms to make bidirectional data communication more reliable. *Quick Exchange (QE)* helps packets traveling in opposite directions, *Fast Forward (FF)* tries to reduce self-contention for packets with the same direction. Both are extensions to 802.11's RTS/CTS mechanism.

QE allows to exchange two packets in opposite directions by using only one exchange of RTS/CTS information. By adding an extra duration header to the first data transmission the network allocation vector of all other nodes in range is extended appropriately. A packet in the opposite direction with a piggybacked ACK can then be sent directly, i. e. without a second RTS/CTS. After both transmissions the original sender completes the procedure with an additional ACK.

FF speeds up the forwarding of a packet in the downstream direction. Like QE there is only one exchange of RTS/CTS information. But here the ACK is piggybacked with a new RTS packet for forwarding. This way, packets are forwarded faster over multiple hops to avoid self contention with other packets of the same flow. The use of this technique is restricted probabilistically because too frequent usage can lead to flow unfairness.

In simulation studies, the authors see a performance gain with their techniques, but also some problems in the interaction of FF with TCP: FF causes a high RTT variance, leading to suboptimal TCP performance.

Against the background of shared medium effects Zhai at al. in [ZF06] propose four mechanisms to reduce the impact that inter-flow and intra-flow contention have on the throughput and fairness in MANETs. They call their approach **Optimum Packet scheduling for Each Traffic flow (OPET)**. For reliable end-to-end communication they combine the proposed mechanisms with standard TCP.

The authors propose to assign a higher priority for the medium access to a node that has just received a packet. This is to give "downstream" packet transmissions a higher priority and thus to alleviate intra-flow contention. This has a similar background as the previously described Fast Forwarding mechanism. In addition, a hop-by-hop backward-pressure scheme keeps upstream nodes from sending further packets until the previous ones have been forwarded. This mechanism is tightly coupled to the 802.11 RTS/CTS mechanism, allowing the receiving node to send a "negative CTS" (NCTS) in order to signal that it is not willing to receive another packet of a certain flow. The upstream node then has to wait until its next hop explicitly gives the permission to continue.

The third mechanism limits the number of packets that source nodes can inject into their local queues, in order to prevent the sources from consuming too much local bandwidth for themselves. Finally, a flow-based round-robin scheme keeps, e. g., a single hop flow from occupying the medium excessively.

Like Fu et al. had done before with ADTFRC (see section A.2), Li et al. propose an adaption of TFRC to MANET requirements called **RE TFRC** [LLA⁺04]. RE TFRC is intended to alleviate negative MAC layer influences on the TFRC rate control mechanism. The authors show that unmodified TFRC produces a high load, beyond the MAC's saturation point. They introduce a rate estimation (RE) algorithm into TFRC. It uses a model for the round trip time to derive the loss rate equivalent to the load saturating the MAC layer capacity. The rate estimation algorithm is used to constrain the rate of TFRC.

The authors regard RTS/CTS-induced congestion as a major reason for the performance drop in an overload situation. Their rate estimation algorithm is based on a wireless multihop network model that is essentially equivalent to a collection of independent single hop wireless networks. This model is used to derive the optimal round trip time upon MAC saturation, which in turn serves as a basis for constraining the TFRC rate below this point.

RE TFRC's approach to tackle the problems of transport protocols in wireless multihop networks by refining the modeling seems promising. However, the question arises whether the approach will work in more complex scenarios than evaluated by the authors, e. g., when interactions between multiple flows occur. It is also interesting to compare the findings in this paper to those of Chen and Nahrstedt in [CN04]. They show fundamental problems of equation-based congestion control in MANETs. Whereas in RE TFRC it is observed that the offered load is beyond the MAC's saturation point, Chen and Nahrstedt report a too conservative behavior of TFRC.

A.4 Handling ACK Traffic

Because of the shared medium, packets using the same route—or spatially close routes—in opposite directions severely affect each other. A very prominent example for this situation is the end-to-end acknowledgment traffic generated by transport protocols, causing intra-flow contention between data packets and acknowledgment packets traveling in opposite directions.

The question arises how the amount of ACK traffic or at least its negative impact on the performance of the forward channel can be minimized. This is closely related to the effects caused by the shared medium in general. Consequently there is some overlap with the previous section, and some of the approaches described there also consider the interplay between oppositely-directed data and ACK traffic. The work described in this section focuses solely on the acknowledgment traffic.

Altman and Jiménez in their **dynamic delayed ACK** approach follow this direction in [AJ03]. Using delayed acknowledgments (DACK) in MANETs has been proposed before, e. g., in [DB01, SM01]. Here, the authors extend the idea beyond standard DACK's combination of only two consecutive ACKs (see RFC 1122 [Bra89]). In the Dynamic Delayed ACK scheme, only after a given number d of segments or after a certain, fixed timeout an acknowledgment packet is sent.

For $d = 2$ the authors observe significant performance improvements in their ns-2 simulations, which increase further for higher values of d of 3 or 4. However, these values might be problematic when TCP operates at a small window size. Therefore the authors propose to use a dynamic delayed ACK with d growing with an increasing packet sequence number, up to $d = 4$. Once this limit is reached, d is never decremented again. They state that a value of d depending on the current window size would probably lead to better results, but they do not want to introduce the additional changes required in order to make this information available at the receiver. Simulations demonstrate the performance gains possible with this approach, but are performed only with single flows in static networks.

De Oliveira and Braun in [dOB07] identify some drawbacks in Altman and Jiménez' scheme. They criticize the lack of adaptability to changing medium conditions. They design a new scheme called **dynamic adaptive acknowledgment** by applying concepts proposed in RFC 2581 (TCP Congestion Control) [APS99] to TCP in MANETs, and by a dynamic ACK timeout which is calculated based on the packet inter-arrival times at the receiver.

The concepts from RFC 2581 comprise an immediate acknowledgment upon out-of-order packets or packets filling a gap at the receiver. For the timeout a sliding average over the packet inter-arrival times is maintained. The authors argue that it is reasonable to wait at least for the time until the second next packet should arrive before a timeout occurs. If this packet is in-order and just delayed, everything is fine. Otherwise, if a packet is missing, the next packet will be out-of-order and will thus trigger an immediate acknowledgment.

In addition to these mechanisms, the parameter d is also changed dynamically. It grows additively up to a maximum value of 4. On out-of-order or gap-filling packets it is reduced to the Delayed ACK standard value of 2. Additionally, the congestion window is limited to a maximum of 4 segments.

The positive results obtained with delayed acknowledgment schemes in general are very interesting, because these techniques can be combined with many other approaches quite well, and are a promising way of reducing the number of packets and thus the contention for the shared medium. It could also be considered to use similar techniques for aggregating the feedback from the receiver to the sender in other transport protocols.

An in-network method for dealing better with acknowledgment traffic, the **preferred ACK retransmission**, is proposed in [SM03] by Sugano et al. As a basis for their system, they suggest to combine two well-known approaches from the literature, namely ELFN messages and the DACK option for TCP on the Flexible Radio Network (FRN). FRN is a commercially available MANET system by Fuji Electric. It is not based on the IEEE 802.11 standard and uses fixed time slots on the medium. On this foundation, they put forward an additional improvement, intended to avoid repeated collisions of ACKs with data packets from the same stream. Their idea is to give acknowledgment packets a higher priority on the medium, by assigning them a shorter MAC retransmission interval.

Also tailored for the FRN is a technique proposed by Yuki et al. in [YYS+04]. They use a mechanism in intermediate nodes for **combining oppositely-directed TCP data and ACK packets** into one common frame.

Their main idea is to avoid using a whole FRN time slot for a very short ACK frame and at the same time to reduce the probability of packet collisions by combining a data and an ACK packet into a common frame if the transmitting node has packets of both kinds in its queue. This frame has two destination addresses, one for each of the two parts. One of the two next hop nodes delays forwarding the packet for one time slot, in order to avoid a collision when both parts are forwarded further.

The authors also discuss the applicability of their mechanism to generic ad-hoc networks. Although no results for other network systems than FRN are given, they consider the technique also useful in other environments, given that the composition and decomposition of frames containing multiple packets is feasible.

A.5 Limiting TCP's Packet Output

Fu et al. showed in [FZL⁺03] that a small TCP congestion window can have beneficial effects on the performance in mobile ad-hoc networks (see section A.3). Following that, a number of approaches have been proposed that exploit more or less directly this effect.

In [CXN03] Chen et al. take up Fu et al.'s observation. On this basis they establish a **dynamic congestion window limit** based on the broadcast characteristics of the wireless medium.

They argue that the congestion window limit (CWL) depends on the bandwidth-delay product (BDP) of the connection, and they show that the BDP cannot exceed the round-trip hop-count (RTHC) in a wireless multihop network. For the IEEE 802.11 MAC they give an even tighter bound of $\frac{RTHC}{5}$.

They use DSR as a path-aware routing protocol to get information about the path length at the source node. This allows for setting the CWL dynamically depending on the path length of the connection.

To demonstrate the performance gain of their scheme ns-2 simulations have been conducted to compare it to TCP Reno with an unbounded congestion window. However, it should be noted that they also changed the maximum retransmission timeout of TCP in their simulations, setting it to 2 s as opposed to the 240 s given in RFC 1122 [Bra89]. This might affect the simulation results.

This modification of the maximum retransmission timer is criticized by Papanastasiou and Ould-Khaoua in [POK04]. They also propose their own scheme, called **Slow Congestion Avoidance (SCA)**. Their approach is to limit TCP's window growth rate to a level below the standard of one segment per RTT. This is intended to reduce the number of packets in the network without putting hard constraints on the maximum window size like in the dynamic CWL scheme.

The SCA modification of the TCP window increment mechanism increases the window size by one segment after a given number of round trip times with successful acknowledgment receptions.

SCA seems to be an interesting approach, showing ways to deal with a MANET's shared channel properties without the need to use cross-layer information in the transport protocol. Further investigation could be necessary in order to examine the properties of SCA, especially under different traffic loads.

An adaption of the TCP behavior to mobile ad-hoc networks by Nahm et al. [NHK05] is similar to the ideas realized in SCA. They also propose to reduce the rate of the congestion window growth of TCP. They call their scheme **fractional window increment (FeW)**.

The reasoning for the growth rate limitation in FeW is the observation that TCP generally operates at a high loss rate in networks with a low bandwidth-delay product. But because losses in congested wireless networks are usually link layer losses rather than queue overflows, these losses also influence routing—the routing protocol will often assume a lost link. The FeW approach is to change TCP's operational range, in order to achieve a generally lower loss rate.

A mathematical analysis based on the TCP-friendly steady state throughput equation is used to deduce that such a shift of the operational range can be achieved by incrementing TCP's congestion window slower than in standard TCP. In practice this leads to a scheme with a non-integer increment in the window size per RTT. It is equivalent to SCA's window size increment only every n round trip times.

The promising results of SCA and FeW should definitely be taken into account by other researchers working on modified TCP variants for MANETs. However, it is not yet clear to which extent short connections with only relatively small amounts of data might suffer from the slower congestion window growth and the resulting slower convergence.

Yang et al. looked at MANETs that are connected to a wired backbone network. There, unmodified TCP exhibits severe unfairness. But in [YSY03] they also point out that simply reducing the congestion window size will severely degrade the performance in such a setup: the small window size only allows for a small number of packets present in the network in parallel, and this affects the wired part of the network, too. As a remedy they propose a **non-work-conserving scheduling** mechanism for the wireless interfaces.

In their scheme, after transmitting a data packet on the wireless medium a timer is set. Before this timer expires no other data packet is allowed to be sent by the same node. The run time of the timer increases with a higher output data rate of the queue in the recent past. Therefore more aggressive nodes' queues are slowed down more. The authors observe greatly improved fairness in their simulations, but at the same time a significant throughput deterioration.

While the approaches described so far directly modify TCP's congestion window size, the non-work-conserving scheduling reduces the rate at which packets are allowed to be forwarded. This happens at every intermediate node and per interface queue. There are also proposals which add an additional rate based control mechanism to the output of the transport layer only in the source nodes.

In general the impact of those approaches is similar to the window size limitation schemes—the number of packets per time unit and per TCP sender is limited. However, rate-based mechanisms might be able to achieve additional benefits by reducing the inherent burstiness of TCP traffic and distributing packet transmissions more evenly. This might significantly reduce intra-flow contention. On the other hand, because multiple, stacked rate limitation and congestion control mechanisms are used the question of possible feedback-loop effects arises.

The **Rate-Based Congestion Control (RBCC)** mechanism proposed by Zhai et al. in [ZCF05] adds a leaky bucket mechanism beyond TCP's window-based rate control. In RBCC, a feedback field is added to the header, which is used by all intermediate nodes to provide feedback on the allowed maximum rate of the flow.

Each node on the route observes its local *channel busyness ratio*, defined as the fraction of time the medium is locally non-idle. It is used to modify the feedback field in the packets passing by, in order to inform the source about the sustainable rate. The intermediate nodes aim to distribute the capacity fair between the flows. For that reason, state describing the flows passing through the local node is maintained, and an AIMD mechanism is used to converge to fairness.

Kliazovich et al. measure in **Cross-layer congestion control (C^3TCP)** [KG06] the bandwidth and delay within an end-to-end link by cumulating intermediate hops' measurements. Like in RBCC a feedback field is added to the link layer header. The measurements are gathered hop by hop from the intermediate nodes and stored in the feedback field. When an ACK is generated at the destination node the feedback included in the corresponding data packet is repeated and thus transmitted back to the sender. There it is used by an additional module—located beyond TCP in the protocol stack—to modify the receiver advertised window (rwnd) field in the ACK. In this field the receiver can limit the sender's window size. Its normal purpose is flow control, in C^3TCP it is used to limit the window size of the sender dynamically based on the measurements. In order to keep the TCP implementation unmodified, all C^3TCP logic is contained within the additional protocol module.

While RBCC and C^3TCP rely on the participation of intermediate nodes, ElRakabawy et al. propose a pure end-to-end approach [EKL05]. They suggest to pace the packets allowed to be sent out by the congestion window adaptively. Their approach is called **TCP with Adaptive Pacing (TCP-AP)**. Like RBCC, TCP-AP is a hybrid between a window-based and a rate-based approach, adding rate-based mechanisms to TCP in order to avoid large bursts of packets.

As a metric to be evaluated in order to configure the pacing the *4-hop propagation delay* is defined. It describes the time between the transmission of a packet by the TCP source node and its reception by the node four hops downstream. Since TCP-AP is a pure end-to-end protocol, the 4-hop propagation delay cannot be measured directly. Instead, it is estimated using the RTT of the packets. The 4-hop propagation delay is chosen by the authors because a transmission currently in progress is assumed to interfere within a range of four hops, a number matching the common modeling assumptions as used, e. g., in the ns-2 simulator.

In addition to the 4-hop propagation delay, the coefficient of variation of RTT samples is proposed as a metric. Its purpose is to measure the degree of contention along the path. In combination with the 4-hop propagation delay it is used to establish a minimum time between two successive packet transmissions.

A.6 Alternative Protocol Designs

A variety of wireless peculiarities have been shown to be detrimental to TCP's end-to-end way of performing congestion control in mobile ad-hoc networks up to now. Consequently some researchers do not just try to trim TCP to perform better by adjusting the protocol's behavior. Instead they develop new reliable transmission protocols that are specifically tailored to cope with the characteristics of MANETs. While the authors of these approaches report to have obtain a broad range of improvements, in these approaches this necessarily comes at the cost of TCP compatibility. Moreover, most approaches are also limited to "clean" environments where no other transport protocols are used.

However, since MANETs can often be expected to be rather small, closed environments, such constraints can be perfectly reasonable. Additionally, it seems that it will be absolutely necessary to rely on completely different queuing and congestion control paradigms than those used by TCP in networks with media properties like those of wireless multihop networks. At least there are results that show very fundamental problems of TCP-like mechanisms in the presence of wireless interference [RK06].

The first representative of the alternative protocol category is the **EXACT** protocol by Chen et al. [CN02, CNV04]. EXACT is rate based and is supported by the network itself, i. e., by the intermediate nodes. These nodes have dedicated state variables for all flows passing through them. All nodes determine their current bandwidth to their neighbors and calculate local fair bandwidth shares for all flows.

Explicit rate information is inserted into all passing packets by the intermediate nodes to transmit the minimum bandwidth at the bottleneck to the receiver of the flow. Each node checks whether the rate it can supply for the flow of a packet it processes is lower than the rate currently specified in the packet header. In this case the lower rate is written into the header before the packet is forwarded. Thus the bottleneck rate is reported in the end.

This mechanism is used twice, i. e. on two different header fields. One field contains the current rate of the sender and another one the rate requested by the sending application. On the one hand with this procedure it is possible for the intermediate routers not to give a flow more bandwidth than it needs, and on the other hand the sender is notified when it is allowed to increase its rate above the current level.

A *safety window* prevents the sender from overloading the network in case of a route failure. A sender is not allowed to have more unacknowledged packets underway than the size of the safety window.

EXACT may be used reliably (TCP-EXACT) or unreliably (UDP-EXACT). There are no retransmission timers, instead a SACK scheme with strictly monotonous increasing sequence numbers is used: when a segment not acknowledged by the receiver is too far apart (in terms of this sequence number) from the highest acknowledged segment, it is retransmitted.

Some limitations on EXACT's practical usage and scalability might be imposed by the fact that it requires explicit state information for each flow in each intermediate node.

Another protocol offers some similar properties to those of EXACT. It is also rate based and network supported. Sundaresan et al. tailored it to the specific needs in MANETs and called it **Ad-hoc Transport Protocol (ATP)** [SAHS03]. It does not use retransmission timeouts, strictly separates congestion control and reliability mechanisms and requires only limited feedback from the receiver. In contrast to EXACT, ATP does not require any flow-specific state variables in the intermediate nodes. All nodes calculate an exponential average of the delay of all packets passing through them. This delay consists of the time a packet had to wait in the node's local queue and of the time to wait for a free medium before it could be transmitted. These values are independent of the flows the packets belongs to.

Like the rate information in EXACT, the current delay value is piggybacked onto forwarded data packets if it is worse than the information currently in the packet's header. This way, the maximum delay over the packet's path is communicated to the receiver. The receiver aggregates this information and sends it back to the sender. Based on this information the sender can adapt its rate.

To find a good rate at the start of a new connection a probe packet is sent along the route collecting information from the intermediate nodes about the current state of the network. For the acknowledgments of data packets a selective ACK scheme is employed in ATP. It uses large SACK blocks and therefore requires only few feedback packets.

A second approach which is also called ATP, the **Application controlled Transport Protocol** by Liu and Singh [LS99], is based on the observation that TCP's throughput in MANETs is very low, while UDP achieves reasonable throughput, but suffers from a high packet loss rate. ATP is meant to be somewhere in between TCP and UDP—UDP with optional packet delivery status feedback.

The protocol supports packet acknowledgments, feedback is given to the application whether an acknowledgment for a given packet has arrived or not. An application using ATP is expected to do retransmissions on its own, if they are necessary.

Leaving the decision upon whether retransmissions are necessary or not to the applications is an interesting approach to reducing the number of retransmitted packets. However, no other transport layer components are implemented, especially congestion control would have to be provided by the application itself. Therefore ATP's approach alone can probably not be sufficient to save a MANET from severe congestion problems.

175

Not a complete redesign, but a variation of the eXplicit Control Protocol (XCP) [KHR02] for wired networks with high bandwidth-delay product is the **Wireless eXplicit Congestion control Protocol (WXCP)** by Su and Gross [SG05]. Although it shares some fundamental concepts with TCP, XCP is not compatible with standard TCP. WXCP uses explicit feedback from within the network and multiple congestion metrics. These are evaluated at the intermediate nodes, in order to avoid the necessity of probing for the highest available bandwidth.

The metrics used in each WXCP-enabled network node are the locally available bandwidth, the length of the local interface queue and the average number of required link layer retransmissions. The latter is specifically meant to help detecting self-interference within a flow, that is, packets belonging to the same flow contending for medium access in the same collision domain. The aggregate feedback is a function of the three metrics, weighting their relative influence.

Congestion and fairness control decisions are made separately in WXCP. The fairness controller tries to achieve time fairness instead of throughput fairness among flows, since throughputs of different links may not be the same in wireless networks. Time fairness, guaranteeing all flows equal medium access time as opposed to equal throughputs, is thus regarded as a better fairness metric by the authors.

WXCP is a window-based approach with some rate-based elements in it. The sender may switch from the window-based default to a slow rate-based control mechanism, if otherwise no further packets were allowed to be transmitted—due to a small congestion window and missing ACKs or duplicate ACKs. This is called the *discovery state*. It allows the sender to continuously examine the current packet loss pattern.

Additionally a pacing mechanism introduces rate-based ideas into WXCP. If the number of packets allowed to be sent out by the window mechanism exceeds a certain limit, the packets are paced to be evenly distributed over an RTT interval.

A transport protocol for mobile ad-hoc networks developed from scratch by Anastasi et al. is **TPA** [AACP05]. Its congestion control mechanism is inspired by TCP, but designed to minimize the number of required packet retransmissions.

Packets are transmitted in *blocks* using a window-based scheme. A fixed number of packets is grouped into a block and transmitted reliably to the destination before any packet of the next block is transmitted. Packet retransmissions are not performed before every packet of a block has been transmitted once—thus a block is transmitted in several rounds: first every packet is transmitted once, then not yet acknowledged packets of this block are retransmitted until every packet of the block has been delivered and acknowledged.

If an ELFN mechanism is available, TPA can make use of it and enters a freeze state upon route failures, decreasing the window size to one. If ELFN is not available, TPA detects route failures by a number of consecutive timeouts. Like TCP it uses an estimate of the RTT to set the retransmission timeout. In case of route changes, new RTT values are given a greater weight in the sliding average in order to speed up the convergence against a correct new RTT measure.

For congestion control TPA uses a window mechanism with a tightly limited maximum window size. Actually, only two different cwnd values are used: a "large" window of 2 or 3 segments during normal operation and the minimum value of 1 when congestion is detected.

TPA shows that even a quite simple end-to-end protocol without additional intelligence in the intermediate nodes has the potential to increase throughput in comparison to TCP. However, it is not yet clear if these benefits can be maintained in more complex, dynamic scenarios. Additionally, for time critical applications the higher latency introduced by the protocol might be a problem.

Appendix B

Error Bounds and Consistency of MLE Time Synchronization

In Chapter 6 we have proposed a solution to the problem of generating a time-consistent global log file out of a set of local log files from a number of network nodes. The local clocks of these nodes are inaccurate and thus introduce errors in the local timestamps. In our approach we consider a network with a local broadcast medium, i. e., one where all or parts of the nodes observe certain events at the same time instant. These common events are used as a kind of anchor points for a maximum likelihood estimation of the clock error and of additional delays occurring in each node upon logging the events. The latter are called the timestamping delays. This maximum likelihood estimator leads to a big linear program (LP).

Our proposed approach is able to estimate and compensate linear-affine clock deviations. Here we consider a simplified variant that ignores deviations of the clocks' rates, such that the only deviations of the clocks are constant offsets. We prove two desirable properties for this version of the estimator. First, we show that tight error bounds on the estimation error hold under the assumption of a bounded timestamping delay. In particular this means that the algorithm does not amplify errors. Furthermore, we show that our estimator is consistent. This means that for increasing data set sizes the estimate converges (in probability) to the true values of the estimated features. It thus supports the intuition that the estimate improves for a larger amount of observed and logged events in the nodes.

B.1 The Simplified Estimator

For the most part we adopt the notation from Chapter 6 here. The set of nodes is denoted by J and the set of events by I. R is the relation of node-event pairs for which a timestamp has been recorded, i.e., $(i,j) \in R \subseteq I \times J$ if and only if i has been observed and timestamped by j.

The employed clock model assumes that for each node j there is an offset o_j and a rate r_j such that j's clock maps the real time t to the local time $C_j(t) = r_j t + o_j$. In addition to the clock deviations there are timestamping errors which are assumed to be independent and exponentially distributed with parameter λ. This means that for the full-featured maximum likelihood estimator the time recorded for event i at "real" time T_i by node j is $r_j(T_i + d_{i,j}) + o_j$ where $d_{i,j}$ is a random variable modelling the timestamping delay.

Here we consider a simplified version of the above. We assume that the clocks run (approximately) at the correct rate, i.e., we set $\forall j \in J : r_j = 1$. Under this assumption the recorded time for a node-event pair $(i,j) \in R$ becomes $T_i + d_{i,j} + o_j$. Thus the simplified maximum likelihood estimator, in analogy to the full-featured version, is the solution to the following problem:

$$\text{minimize } L = \prod_{(i,j)\in R} \lambda e^{-\lambda\left(t_{i,j} - \widehat{o}_j - \widehat{T}_i\right)}$$

$$\text{subject to } \forall (i,j) \in R : \widehat{d}_{i,j} = t_{i,j} - \widehat{o}_j - \widehat{T}_i \geq 0.$$

Here, $t_{i,j}$ denotes the local time when node j has recoded event i. $\widehat{o}_j$ and $\widehat{T}_i$ are the estimates of o_j and T_i, respectively.

As shown in Chapter 6, the optimal solution is independent of λ and equivalent to solving

$$\text{minimize } k(L) = \sum_{(i,j)\in R} \left(t_{i,j} - \widehat{o}_j - \widehat{T}_i\right) = \sum_{(i,j)\in R} \widehat{d}_{i,j}$$

under the same constraints as above.

Let an optimal solution of this LP, consisting of the estimates $\widehat{T}_i$, $\widehat{o}_j$ and $\widehat{d}_{i,j}$, be denoted by $\mathcal{S}$.

To simplify the following, we use notations for the set of all nodes observing a certain event and for all the events observed by a given node:

$$\forall i \in I : R_i := \{j \in J | (i, j) \in R\}$$
$$\forall j \in J : R^j := \{i \in I | (i, j) \in R\}$$

B.2 Error Bounds

Our intention in this section is to establish an upper bound on the error of the maximum likelihood estimator, i.e., the maximum difference between estimated and real event times. In order to do so we make two additional assumptions. The first one guarantees network connectivity, the second one establishes an upper bound on the timestamping delay.

Before we introduce the assumptions it is necessary to recall one important fact from Chapter 6, which we call the *offset ambiguity*: for any offset $\tau \in \mathbb{R}$, the estimates $\widehat{T}_i$, $\widehat{o}_j$ and the estimates $\widehat{T}_i'$, $\widehat{o}_j'$ with

$$\forall i \in I : \widehat{T}_i' = \widehat{T}_i + \tau \qquad \forall j \in J : \widehat{o}_j' = \widehat{o}_j - \tau$$

fit the same set of measurements equally well.

From the offset ambiguity it is easy to see that there is also no way to estimate all the relative times within an experiment if the network is partitioned. If there are no anchor points between two sets of nodes, there will be an ambiguity of the offset between these partitions within the experiment. Thus, in order to get a bounded maximum estimation error we need to assume network connectivity. The *connectivity assumption* says that the network nodes do not fall into disjoint partitions, between which no anchor points at all exist. Formally this is

$$\forall P_1, P_2 \subseteq J, \ P_1, P_2 \neq \emptyset :$$
$$(P_1 \cup P_2 = J \ \wedge \ P_1 \cap P_2 = \emptyset \implies \exists j_1 \in P_1, j_2 \in P_2 : R^{j_1} \cap R^{j_2} \neq \emptyset).$$

Under this assumption we will prove that

$$\forall j_1, j_2 \in J : |(o_{j_1} - o_{j_2}) - (\widehat{o}_{j_1} - \widehat{o}_{j_2})| \leq (|J| - 1) \cdot D$$

and

$$\forall i_1, i_2 \in I : \left|(T_{i_1} - T_{i_2}) - \left(\widehat{T}_{i_1} - \widehat{T}_{i_2}\right)\right| \leq |J| \cdot D$$

if $D \in \mathbb{R}^+$ is an upper bound for the delays, i. e.,

$$\forall (i,j) \in R : d_{i,j} \leq D.$$

Note that the bounds are on the difference between two estimation errors because of the offset ambiguity.

The following proof does not exploit the exponential distribution of the delays. Thus, independent of the derivation of the estimator, the proof shows that if there is an upper bound for the timestamping delays the estimates are close to the real values, regardless of the real distribution of the delays within $[0, D]$. Note that assuming the existence of such an upper bound does not limit the practical applicability of the results given here: for any given experiment, the set of observations R is finite, and thus there will always be a maximum delay.

For the proof we introduce some additional terminology.

Definition Let $j, j' \in J, i \in I$. We call i a *common event* of j and j' iff $\{(i, j_1), (i, j_2)\} \subseteq R$, and we call i a *connecting event* from j to j' iff i is a common event of j and j' and the timestamping delay of i in j is estimated as zero, i. e., $\widehat{d}_{i,j} = 0$.

As will soon become clear, a significant part of the timestamping delay estimates are zero. We now construct a directed graph $G := (J, E)$ with

$$E := \left\{(j, j') \in J^2 \mid \exists i \in I : i \text{ is a connecting event from } j \text{ to } j'\right\}.$$

The graph G plays a central role in our proof. In the following two lemmas we point out some properties of G.

Lemma B.1. *Let $j_A, j_B \in J$. Then there exists a directed path $(j_A, \ldots, j_B)$ in G.*

Proof The set of nodes J can be divided into two disjoint subsets:

$$\begin{aligned} J_1 &:= \{j \in J \mid \text{there exists no directed path } (j, \ldots, j_B) \text{ in } G\} \\ J_2 &:= J \setminus J_1. \end{aligned}$$

In the following, we will show that J_1 is empty. Let $\bar{I}$ be the events occurring in both J_1 and J_2:

$$\bar{I} := \{i \in I \mid \exists j_1 \in J_1, j_2 \in J_2 : \{(i,j_1),(i,j_2)\} \subseteq R\}.$$

Let $j_1 \in J_1, j_2 \in J_2$. Then there is no connecting event from j_1 to j_2. Otherwise a path from j_1 to j_B could be constructed by concatenation of (j_1, j_2) and $(j_2, \ldots, j_B)$, which is a contradiction to $j_1 \in J_1$. Thus, we have

$$\exists \epsilon > 0 : \forall (i,j) \in R \cap (\bar{I} \times J_1) : \widehat{d}_{i,j} \geq \epsilon.$$

Let now

$$
\begin{aligned}
I_1 &:= \{i \in I \mid \nexists j \in J_2 : (i,j) \in R\} \\
I_2 &:= \{i \in I \mid \nexists j \in J_1 : (i,j) \in R\}.
\end{aligned}
$$

With these definitions $I = I_1 \cup I_2 \cup \bar{I}$ holds, and $I_1, I_2, \bar{I}$ are pairwise disjoint. Now a new solution $\mathcal{S}'$ of the LP can be constructed:

$$
\begin{aligned}
\widehat{T}'_i &:= \widehat{T}_i - \begin{cases} \epsilon & \text{if } i \in I_1 \\ 0 & \text{otherwise} \end{cases} \\[2mm]
\widehat{o}'_j &:= \widehat{o}_j + \begin{cases} \epsilon & \text{if } j \in J_1 \\ 0 & \text{otherwise} \end{cases} \\[2mm]
\widehat{d}'_{i,j} &:= \widehat{d}_{i,j} - \begin{cases} \epsilon & \text{if } i \in \bar{I} \wedge j \in J_1 \\ 0 & \text{otherwise.} \end{cases}
\end{aligned}
$$

It can easily be verified that all the LP constraints hold for $\mathcal{S}'$ since they hold for $\mathcal{S}$. For $\mathcal{S}'$ we have

$$\sum_{(i,j) \in R} \widehat{d}'_{i,j} = \sum_{(i,j) \in R} \widehat{d}_{i,j} - \left| R \cap (\bar{I} \times J_1) \right| \cdot \epsilon.$$

Since $\mathcal{S}$ is optimal and $\epsilon > 0$ we have $\left| R \cap (\bar{I} \times J_1) \right| = 0$. Therefore, $\bar{I} = \emptyset$ or $J_1 = \emptyset$, due to the definitions of J_1 and $\bar{I}$ above.

$\bar{I}$ is not empty if J_1 is not empty due to the connectivity assumption. Therefore, $J_1 = \emptyset$ and $j_A \in J_2$. Thus, there exists a directed path $(j_A, \ldots, j_B)$ in G. $\qquad \square$

Lemma B.2. *Let* $(j_0, \ldots, j_n) \in J^{n+1}$ *be a directed path in* G. *Then*

$$(o_{j_0} - o_{j_n}) - (\widehat{o}_{j_0} - \widehat{o}_{j_n}) \leq nD.$$

Proof We prove this by induction.

Let $n = 1$. Let i be a connecting event from j_0 to j_1. Such an event exists due to the construction of G. From the LP we have

$$T_i + o_{j_0} + d_{i,j_0} = \widehat{T}_i + \widehat{o}_{j_0} + \widehat{d}_{i,j_0} \tag{B.1}$$

$$T_i + o_{j_1} + d_{i,j_1} = \widehat{T}_i + \widehat{o}_{j_1} + \widehat{d}_{i,j_1}. \tag{B.2}$$

The difference between (B.1) and (B.2) yields

$$(o_{j_0} - o_{j_1}) + (d_{i,j_0} - d_{i,j_1}) = (\widehat{o}_{j_0} - \widehat{o}_{j_1}) + (\widehat{d}_{i,j_0} - \widehat{d}_{i,j_1})$$

$$\Longleftrightarrow \quad (o_{j_0} - o_{j_1}) - (\widehat{o}_{j_0} - \widehat{o}_{j_1}) = \underbrace{\underbrace{(\widehat{d}_{i,j_0}}_{=0} - \underbrace{\widehat{d}_{i,j_1})}_{\geq 0}}_{\leq 0} - \underbrace{(\underbrace{d_{i,j_0}}_{\in [0,D]} - \underbrace{d_{i,j_1}}_{\in [0,D]})}_{\in [-D,D]}$$

$$\Longleftrightarrow \quad (o_{j_0} - o_{j_1}) - (\widehat{o}_{j_0} - \widehat{o}_{j_1}) \leq D.$$

For the induction step we have:

$$(o_{j_0} - o_{j_{n-1}}) - (\widehat{o}_{j_0} - \widehat{o}_{j_{n-1}}) \leq (n-1)D \tag{B.3}$$

$$(o_{j_{n-1}} - o_{j_n}) - (\widehat{o}_{j_{n-1}} - \widehat{o}_{j_n}) \leq D. \tag{B.4}$$

(B.4) can be constructed like above. Addition of (B.3) and (B.4) gives us

$$(o_{j_0} - o_{j_{n-1}}) - (\widehat{o}_{j_0} - \widehat{o}_{j_{n-1}}) + (o_{j_{n-1}} - o_{j_n}) - (\widehat{o}_{j_{n-1}} - \widehat{o}_{j_n}) \leq (n-1)D + D$$

$$\Longleftrightarrow \quad (o_{j_0} - o_{j_n}) - (\widehat{o}_{j_0} - \widehat{o}_{j_n}) \leq nD.$$

$\square$

Now that we know that G is connected and that an upper error bound holds for any path in G, we only need to put together these pieces in order to get a bounded error of the offsets for any pair of nodes. While Lemma B.2 establishes only an upper bound for the differences of two offset estimation errors, the fact that G is connected and thus paths in both directions exist can be exploited to bound this difference from below, too.

Theorem B.3. *Let $j_1, j_2 \in J$. Then the following bound holds*

$$|(o_{j_1} - o_{j_2}) - (\widehat{o}_{j_1} - \widehat{o}_{j_2})| \leq (|J| - 1)\, D.$$

Proof According to Lemma B.1 there exists a path from j_1 to j_2 in G. Likewise, there exists a path from j_2 to j_1 in G. Since the number of nodes in G is $|J|$, the maximum length of each of these paths is $|J| - 1$. Thus, we have from Lemma B.2:

$$(o_{j_1} - o_{j_2}) - (\widehat{o}_{j_1} - \widehat{o}_{j_2}) \leq (|J| - 1)\, D$$
$$(o_{j_2} - o_{j_1}) - (\widehat{o}_{j_2} - \widehat{o}_{j_1}) = -((o_{j_1} - o_{j_2}) - (\widehat{o}_{j_1} - \widehat{o}_{j_2})) \leq (|J| - 1)\, D.$$

This immediately gives us the desired result. ☐

We now have an error bound for the offset estimates. From here, it is only a small step to a similar bound for the error in the estimation of the event times. However, in order to give a good bound we need an additional property of the estimator that will be established in the following lemma.

Lemma B.4. *Let $\bar{i} \in I$. Then there exists $\bar{j} \in J : \widehat{d}_{\bar{i},\bar{j}} = 0$.*

Proof We prove this by contradiction. Assume that there is no such $\bar{j}$. Then

$$\exists \epsilon > 0 : \forall (i,j) \in R : i = \bar{i} \Rightarrow d_{i,j} \geq \epsilon.$$

Now we construct an new solution $\mathcal{S}'$ to the LP:

$$\widehat{T}'_i := \widehat{T}_i + \begin{cases} \epsilon & \text{if } i = \bar{i} \\ 0 & \text{otherwise} \end{cases}$$
$$\widehat{o}'_j := \widehat{o}_j$$
$$\widehat{d}'_{i,j} := \widehat{d}_{i,j} - \begin{cases} \epsilon & \text{if } i = \bar{i} \\ 0 & \text{otherwise.} \end{cases}$$

This solution is valid and better than $\mathcal{S}$. Since $\mathcal{S}$ is optimal this is a contradiction. ☐

We can exploit the fact that for any event i a node j exists where $d_{i,j}$ is estimated as zero in the following theorem. The result is the desired error bound for the event time estimates.

Theorem B.5. *Let $i_1, i_2 \in I$. Then the following bound holds*

$$\left|(T_{i_1} - T_{i_2}) - (\widehat{T}_{i_1} - \widehat{T}_{i_2})\right| \le |J| \cdot D.$$

Proof From Lemma B.4 we have

$$\forall k \in \{1, 2\} : \exists j_k \in J : (i_k, j_k) \in R \ \land \ \widehat{d}_{i_k, j_k} = 0.$$

With these j_1, j_2 we have

$$T_{i_1} + o_{j_1} + d_{i_1, j_1} = \widehat{T}_{i_1} + \widehat{o}_{j_1} \tag{B.5}$$

$$T_{i_2} + o_{j_2} + d_{i_2, j_2} = \widehat{T}_{i_2} + \widehat{o}_{j_2}. \tag{B.6}$$

Calculating the difference between (B.5) and (B.6) and some reordering yields

$$(T_{i_1} - T_{i_2}) - (\widehat{T}_{i_1} - \widehat{T}_{i_2}) = \underbrace{(\widehat{o}_{j_1} - \widehat{o}_{j_2}) - (o_{j_1} - o_{j_2})}_{\in[-(|J|-1)D, \ (|J|-1)D]} - \underbrace{(d_{i_1, j_1} - d_{i_2, j_2})}_{\in[-D, D]}.$$

The bounds for $(\widehat{o}_{j_1} - \widehat{o}_{j_2}) - (o_{j_1} - o_{j_2})$ come from theorem B.3.

This result gives us

$$\left|(T_{i_1} - T_{i_2}) - (\widehat{T}_{i_1} - \widehat{T}_{i_2})\right| \le (|J| - 1)D + D = |J| \cdot D,$$

which is the desired bound. □

Now we have bounds on the relative offsets and event time estimates. It remains open whether these bounds are good. To address this question we introduce some additional terms. We then use these to point out an important property of any local broadcast network where distributed log files are recorded. This property in turn will then be used to prove that not only the error bounds are tight, but also that no estimator with better error bounds is possible.

Definition A *receive trace set* $Q = (I, R, (t_{i,j})_{(i,j) \in R})$ for a set of nodes J consists of a set of events I, a relation $R \subseteq I \times J$ and the local timestamps $t_{i,j}$ recorded by the nodes in J for the events they have observed.

For a given set of nodes J, a *scenario* $S = (I, R, (o_j)_{j \in J}, (T_i)_{i \in I}, (d_{i,j})_{(i,j) \in R})$ consists of a set of events I, a relation $R \subseteq I \times J$, and corresponding offsets o_j for all nodes, event times T_i for all events and delays $d_{i,j}$ for each event reception.

From the above definition it is clear that there is exactly one receive trace set for any given scenario. There can, however, well be many possible scenarios for a given receive trace set. The offset ambiguity mentioned at the beginning of this section is a special case of this situation. Note that this is an inherent property of the time synchronization problem as it is considered here, and not specific to our approach.

Definition Two scenarios $S = (I, R, (o_j)_{j \in J}, (T_i)_{i \in I}, (d_{i,j})_{(i,j) \in R})$ and $S' = (I, R, (o'_j)_{j \in J}, (T'_i)_{i \in I}, (d'_{i,j})_{(i,j) \in R})$ are called *indistinguishable* if they share a common set of events I and the same relation R and they result in the same receive trace set, i. e.,

$$\forall (i,j) \in R : o_j + T_i + d_{i,j} = t_{i,j} = o'_j + T'_i + d'_{i,j}.$$

Theorem B.6. *For any set of nodes J, $|J| \geq 1$ there exist two indistinguishable scenarios $S = (I, R, (o_j)_{j \in J}, (T_i)_{i \in I}, (d_{i,j})_{(i,j) \in R})$ and $S' = (I, R, (o'_j)_{j \in J}, (T'_i)_{i \in I}, (d'_{i,j})_{(i,j) \in R})$ and $j_2, j_2 \in J$ such that R fulfills the connectivity assumption and*

$$(o'_{j_1} - o'_{j_2}) - (o_{j_1} - o_{j_2}) = 2(|J| - 1)D.$$

Proof Our proof for this theorem is constructive. Let, for the sake of simplicity and without loss of generality, $J = \{1, \ldots, n\}$. Assume these nodes form a chain-like topology. A total of $n - 1$ events are recorded by the nodes, $I = \{1, \ldots, n - 1\}$. Now we define R as follows

$$(i,j) \in R \iff i \in \{j - 1, j\}.$$

It is easy to see that the connectivity assumption holds for R.

Now let S and S' be defined by

$$\forall i \in I : \quad T_i := (n - i) \cdot D \qquad T'_i := i \cdot D$$
$$\forall j \in J : \quad o_j := j \cdot D \qquad o'_j := (n - j + 1) \cdot D$$
$$\forall (i,j) \in R : \quad d_{i,j} := \begin{cases} 0 & \text{if } i = j - 1 \\ D & \text{if } i = j \end{cases} \qquad d'_{i,j} := \begin{cases} D & \text{if } i = j - 1 \\ 0 & \text{if } i = j. \end{cases}$$

With these definitions it is easy to verify that S and S' are indistinguishable:

$$\forall (i,j) \in R : T_i + o_j + d_{i,j} = T'_i + o'_j + d'_{i,j}.$$

And it also holds that

$$(o'_1 - o'_n) - (o_1 - o_n) = 2(n-1)D.$$

$\square$

Let us now assume that we have two scenarios S and S' like in the above theorem. Assume now we have estimates $\widehat{o}_1$ and $\widehat{o}_n$ of o_1 and o_n that are better than the worst-case result of our maximum likelihood estimator for scenario S. This means that the relative offset estimation error is less than $(|J|-1)D$, which is the bound of our approach according to Theorem B.3. Then in particular

$$(o_1 - o_n) - (\widehat{o}_1 - \widehat{o}_n) > -(|J|-1)D.$$

But we know that

$$(o'_1 - o'_n) - (o_1 - o_n) = 2(n-1)D.$$

By simple addition we get

$$(o'_1 - o'_n) - (\widehat{o}_1 - \widehat{o}_n) > (|J|-1)D.$$

This is worse than the maximum error of our MLE estimator. Since S and S' are indistinguishable, this proves that there cannot be an estimator with lower maximum offset estimation errors: if the estimate is better in case of a receive trace set that resulted from S, it is necessarily worse in case the same receive trace set came from S' and vice versa. Similar results hold for the event time error bounds.

B.3 Consistency

In the last section error bounds for our time synchronization approach have been established. They do not exploit the exponential distribution of the timestamping delay, but rely on an upper bound for the timestamping delay. In this section we will not assume such an upper bound, but we will exploit the exponential distribution of the delays.

Under these premises consistency of the clock offset estimator will be established, which means convergence in probability to the correct offset values for an increasing number of observed events:

$$\forall j \in J : \operatorname*{plim}_{|I| \to \infty} \widehat{o}_j = o_j + x,$$

where $x \in \mathbb{R}$ again comes from the offset ambiguity discussed in the previous section.

We will show the consistency of the simplified MLE under an additional regularity condition, defined as follows.

Definition We say that the *regularity condition* is fulfilled if there exists an undirected, connected graph $G = (J, V)$ and some positive constant β such that

$$\forall \{j_1, j_2\} \in V : E\left[\left| \{i \in I | \{j_1, j_2\} \subseteq R_i\} \right| \right] \geq \beta \cdot |I|.$$

It is valid to assume that G is a tree.

This condition can be seen as a somewhat stronger variant of the connectivity assumption used in the previous section. It is stronger in the sense that it requires an ever-growing number of independent connections between all parts of the network with an increasing total number of observed events, although only probabilistically with respect to the expectancy of their number.

B.3.1 Stochastic Preliminaries

Before we can tackle the main proof some preliminary results from elementary probability theory are necessary. They will be established in the following lemmas.

Lemma B.7. *Let X_1, X_2 be independent, exponentially distributed random variables with parameters λ_1, λ_2. Then $\min \{X_1, X_2\}$ is exponentially distributed with parameter $\lambda_1 + \lambda_2$.*

Proof

$$
\begin{aligned}
P(\min\{X_1, X_2\} \le x) &= P(X_1 \le x \text{ or } X_2 \le x) \\
&= 1 - P(X_1 > x \text{ and } X_2 > x) \\
&= 1 - P(X_1 > x) \cdot P(X_2 > x) \\
&= 1 - \left(\int_x^\infty \lambda_1 e^{-\lambda_1 t} dt \right) \left(\int_x^\infty \lambda_2 e^{-\lambda_2 t} dt \right) \\
&= 1 - \left(e^{-\lambda_1 x} \right) \left(e^{-\lambda_2 x} \right) \\
&= 1 - e^{-(\lambda_1 + \lambda_2)x}.
\end{aligned}
$$

Thus, the minimum of X_1 and X_2 is exponentially distributed with parameter $\lambda_1 + \lambda_2$. □

Lemma B.8. *Let X be a random variable with real values and expected value $E[X] \in \mathbb{R}$. Let $t \in \mathbb{R}$ such that $E[X|X < t]$ and $E[X|X \ge t]$ exist. Then*

$$
E[X] = P(X < t) \cdot E[X|X < t] + P(X \ge t) \cdot E[X|X \ge t].
$$

Proof Assume X is continuous with probability density f.

$$
\begin{aligned}
E[X] &= \int_{-\infty}^{\infty} x \cdot f(x) \, dx \\
&= \int_{-\infty}^{t} x \cdot f(x) \, dx + \int_{t}^{\infty} x \cdot f(x) \, dx \\
&= \int_{-\infty}^{t} x \cdot f(x|X < t) \cdot P(X < t) \, dx \\
&\quad + \int_{t}^{\infty} x \cdot f(x|X \ge t) \cdot P(X \ge t) \, dx \\
&= P(X < t) \cdot E[X|X < t] + P(X \ge t) \cdot E[X|X \ge t].
\end{aligned}
$$

For more general X the proof is analogous. □

Lemma B.9. *Let d be an exponentially distributed random variable with parameter λ. Let $t \in \mathbb{R}^+$. Then*

$$
E[d|d < t] = \frac{1}{\lambda} - \frac{t e^{-\lambda t}}{1 - e^{-\lambda t}}.
$$

Proof From Lemma B.8 and with the memorylessness of the exponential distribution it follows that

$$
\begin{aligned}
E[d|d < t] &= \frac{E[d] - P(d \geq t) \cdot E[d|d \geq t]}{P(d < t)} \\
&= \frac{\frac{1}{\lambda} - e^{-\lambda t}\left(\frac{1}{\lambda} + t\right)}{1 - e^{-\lambda t}} \\
&= \frac{(1 - e^{-\lambda t})\frac{1}{\lambda} - te^{-\lambda t}}{1 - e^{-\lambda t}} \\
&= \frac{1}{\lambda} - \frac{te^{-\lambda t}}{1 - e^{-\lambda t}}.
\end{aligned}
$$

$\square$

Lemma B.10. *Let $d_1, \ldots, d_n$ be independent, exponentially distributed random variables with parameters $\lambda_1, \ldots, \lambda_n$. Let $\Delta_1, \ldots, \Delta_n \in \mathbb{R}$ be given such that $\forall i, 1 \leq i < n :$ $\Delta_{i+1} \geq \Delta_i$.*

With

$$
\forall i, 1 \leq i \leq n : L_i := \sum_{j=1}^{i} \lambda_j
$$

the following equality holds:

$$
\begin{aligned}
E\left[\min_{1 \leq i \leq n}(d_i + \Delta_i)\right] &= \sum_{j=1}^{n-1}\left(\prod_{k=1}^{j-1} e^{-L_k(\Delta_{k+1} - \Delta_k)}\right)\left(1 - e^{-L_j(\Delta_{j+1} - \Delta_j)}\right)\frac{1}{L_j} \\
&\quad + \left(\prod_{k=1}^{n-1} e^{-L_k(\Delta_{k+1} - \Delta_k)}\right)\frac{1}{L_n} + \Delta_1.
\end{aligned}
$$

Proof First observe that if there exists $i, 1 \leq i < n$ such that $\Delta_{i+1} = \Delta_i$, then, with Lemma B.7, this is equivalent to the case where d_i and d_{i+1} are replaced by a single exponentially distributed random variable with parameter $\lambda_i + \lambda_{i+1}$, and the same Δ_i appearing only once. We may thus without loss of generality assume that $\forall i, 1 \leq i < n : \Delta_{i+1} > \Delta_i$.

To simplify the notation of the following, we define

$$
\chi\left[(\Delta_1, \lambda_1), \ldots, (\Delta_n, \lambda_n)\right] := E\left[\min_{1 \leq j \leq n}(d_j + \Delta_j)\right].
$$

This can be generalized for the conditional expected value, i. e., for a condition A, let

$$\chi[(\Delta_1, \lambda_1), \ldots, (\Delta_n, \lambda_n)|A] := E\left[\min_{1 \leq j \leq n}(d_j + \Delta_j)\,\Big|\, A\right].$$

We will now show the assertion by induction over n. The base case $n = 1$ holds:

$$\chi[(\Delta_1, \lambda_1)] = \Delta_1 + \frac{1}{\lambda_1}.$$

For the induction step, we will use the following implication of Lemma B.7:

$$\chi[(\Delta_2, \lambda_1), (\Delta_2, \lambda_2), (\Delta_3, \lambda_3), \ldots, (\Delta_n, \lambda_n)]$$
$$= E\left[\min\{d_1 + \Delta_2, d_2 + \Delta_2, d_3 + \Delta_3, \ldots, d_n + \Delta_n\}\right]$$
$$= E\left[\min\{\min\{d_1 + \Delta_2, d_2 + \Delta_2\}, d_3 + \Delta_3, \ldots, d_n + \Delta_n\}\right]$$
$$= \chi[(\Delta_2, \lambda_1 + \lambda_2), (\Delta_3, \lambda_3), \ldots, (\Delta_n, \lambda_n)].$$

With $\Delta_1 \leq \Delta_2$, Lemma B.8 at (a), the memorylessness of the exponential distribution at (b), Lemma B.9 at (c) and the induction hypothesis (with an index shift) at (d), the following holds:

$$\chi[(\Delta_1, \lambda_1), \ldots, (\Delta_{n+1}, \lambda_{n+1})]$$
$$= E\left[\min\{d_1 + \Delta_1, \ldots, d_{n+1} + \Delta_{n+1}\}\right]$$
$$\overset{(a)}{=} P(d_1 + \Delta_1 < \Delta_2) \cdot E\left[\min\{d_1 + \Delta_1, \ldots, d_{n+1} + \Delta_{n+1}\} \mid d_1 + \Delta_1 < \Delta_2\right]$$
$$\quad + P(d_1 + \Delta_1 \geq \Delta_2) \cdot E\left[\min\{d_1 + \Delta_1, \ldots, d_{n+1} + \Delta_{n+1}\} \mid d_1 + \Delta_1 \geq \Delta_2\right]$$
$$\overset{(b)}{=} P(d_1 + \Delta_1 < \Delta_2) \cdot E\left[d_1 + \Delta_1 \mid d_1 + \Delta_1 < \Delta_2\right]$$
$$\quad + P(d_1 + \Delta_1 \geq \Delta_2) \cdot E\left[\min\{d_1 + \Delta_2, d_2 + \Delta_2, \ldots, d_{n+1} + \Delta_{n+1}\}\right]$$
$$= \left(1 - e^{-\lambda_1(\Delta_2 - \Delta_1)}\right)\left(E[d_1 \mid d_1 < \Delta_2 - \Delta_1] + \Delta_1\right)$$
$$\quad + e^{-\lambda_1(\Delta_2 - \Delta_1)} \cdot \chi[(\Delta_2, \lambda_1), (\Delta_2, \lambda_2), \ldots, (\Delta_{n+1}, \lambda_{n+1})]$$
$$\overset{(c)}{=} \left(1 - e^{-\lambda_1(\Delta_2 - \Delta_1)}\right)\left(\frac{1}{\lambda_1} - \frac{e^{-\lambda_1(\Delta_2 - \Delta_1)}(\Delta_2 - \Delta_1)}{1 - e^{-\lambda_1(\Delta_2 - \Delta_1)}} + \Delta_1\right)$$
$$\quad + e^{-\lambda_1(\Delta_2 - \Delta_1)} \cdot \chi[(\Delta_2, \lambda_1 + \lambda_2), (\Delta_3, \lambda_3), \ldots, (\Delta_{n+1}, \lambda_{n+1})]$$

$$\overset{(d)}{=} \left(1 - e^{-\lambda_1(\Delta_2 - \Delta_1)}\right)\left(\frac{1}{\lambda_1} + \Delta_1\right) - e^{-\lambda_1(\Delta_2 - \Delta_1)}(\Delta_2 - \Delta_1)$$

$$+ e^{-\lambda_1(\Delta_2 - \Delta_1)}\left(\sum_{j=2}^{n}\left(\prod_{k=2}^{j-1} e^{-L_k(\Delta_{k+1} - \Delta_k)}\right)\left(1 - e^{-L_j(\Delta_{j+1} - \Delta_j)}\right)\frac{1}{L_j}\right.$$

$$\left. + \left(\prod_{k=2}^{n} e^{-L_k(\Delta_{k+1} - \Delta_k)}\right)\frac{1}{L_{n+1}} + \Delta_2\right)$$

$$= \left(1 - e^{-L_1(\Delta_2 - \Delta_1)}\right)\left(\frac{1}{L_1} + \Delta_1\right) - e^{-L_1(\Delta_2 - \Delta_1)}(\Delta_2 - \Delta_1)$$

$$+ e^{-L_1(\Delta_2 - \Delta_1)}\left(\sum_{j=2}^{n}\left(\prod_{k=2}^{j-1} e^{-L_k(\Delta_{k+1} - \Delta_k)}\right)\left(1 - e^{-L_j(\Delta_{j+1} - \Delta_j)}\right)\frac{1}{L_j}\right.$$

$$\left. + \left(\prod_{k=2}^{n} e^{-L_k(\Delta_{k+1} - \Delta_k)}\right)\frac{1}{L_{n+1}} + \Delta_2\right)$$

$$= \Delta_1 + \left(1 - e^{-L_1(\Delta_2 - \Delta_1)}\right)\frac{1}{L_1}$$

$$+ \sum_{j=2}^{n}\left(\prod_{k=1}^{j-1} e^{-L_k(\Delta_{k+1} - \Delta_k)}\right)\left(1 - e^{-L_j(\Delta_{j+1} - \Delta_j)}\right)\frac{1}{L_j}$$

$$+ \left(\prod_{k=1}^{n} e^{-L_k(\Delta_{k+1} - \Delta_k)}\right)\frac{1}{L_{n+1}}$$

$$= \Delta_1 + \sum_{j=1}^{n}\left(\prod_{k=1}^{j-1} e^{-L_k(\Delta_{k+1} - \Delta_k)}\right)\left(1 - e^{-L_j(\Delta_{j+1} - \Delta_j)}\right)\frac{1}{L_j}$$

$$+ \left(\prod_{k=1}^{n} e^{-L_k(\Delta_{k+1} - \Delta_k)}\right)\frac{1}{L_{n+1}}.$$

$\square$

B.3.2 Consistency Proof

We will now give the main consistency proof of the simplified maximum likelihood time synchronization estimator. In order to do so, we first formalize the notion of the clock offset estimation error. This definition is actually very trivial; we denote the estimation error by a vector $\Delta = (\Delta_1, \ldots, \Delta_{|J|})^T \in \mathbb{R}^{|J|}$ in the following way.

Definition Let $\forall j \in J : \Delta_j := o_j - \widehat{o}_j$.

In the following, we regard a certain scenario (according to the definition in Section B.2) as given, thus the T_i, o_j and $d_{i,j}$ are fixed. We then consider the estimation error vector

Δ as variable and have a look at the properties of the likelihood function upon a varying estimation error. Our goal is to show that the probability that the likelihood function (regarded as a function of Δ) has its optimum in an arbitrarily small environment around the correct clock offset estimates is arbitrarily high for a sufficing number of observed events.

On our way towards this goal we now introduce a per-event decomposition of the objective function $k(L)$. Certain properties of these event-wise objective function terms form the basis of our proof.

Definition For each event $i \in I$ let f_i be the term added to the objective function $k(L)$ by i for some estimation error vector Δ:

$$f_i(\Delta) := \sum_{j \in R_i} \widehat{d}_{i,j}.$$

The constraints of the LP in our approach are of the form

$$\forall (i,j) \in R : \widehat{d}_{i,j} + \widehat{o}_j + \widehat{T}_i = t_{i,j} = d_{i,j} + o_j + T_i.$$

Thus, the estimated timestamping delays can be expressed as

$$
\begin{aligned}
\widehat{d}_{i,j} &= d_{i,j} + o_j - \widehat{o}_j + T_i - \widehat{T}_i \\
&= d_{i,j} + \Delta_j + (T_i - \widehat{T}_i).
\end{aligned}
$$

Considering that the $d_{i,j}$ and the T_i are given by the scenario it is easy to see that the Δ_j determine the value of the event time estimates $\widehat{T}_i$ chosen by the estimator: all the estimated delays $\widehat{d}_{i,j}$ need to be non-negative and at the same time the sum of all these delay estimates is minimized. Thus, the optimal choice is

$$\widehat{T}_i = T_i + \min_{k \in R_i}(d_{i,k} + \Delta_k).$$

This also follows from the nonnegativity constraints for the $\widehat{d}_{i,j}$ and Lemma B.4.

We can now express the objective function terms f_i in the following way:

$$f_i(\Delta) = \sum_{j \in R_i} \left(d_{i,j} + \Delta_j - \min_{k \in R_i}(d_{i,k} + \Delta_k) \right).$$

This is the formulation that we are going to use throughout the proof. Now we will point out some properties of the objective function terms.

Lemma B.11. *The objective function terms $f_i(\Delta)$ are convex.*

Proof It is well-known that the minimum of concave functions is concave (see, e. g., [HUL91]). Hence, $\min_{k \in R_i}(d_{i,k} + \Delta_k)$ is concave and $-\min_{k \in R_i}(d_{i,k} + \Delta_k)$ is convex. As the sum of convex functions is convex the assertion follows. $\qquad\square$

The next property that we are going to prove is a little more tricky. For simplicity, we will assume that $J = \{1, \ldots, |J|\}$ from now on.

Lemma B.12. *There exists a strictly monotonically increasing function $\alpha : \mathbb{R}_0^+ \to \mathbb{R}_0^+$ such that for each event $i \in I$ and each $\Delta \in \mathbb{R}^{|J|}$ the following holds:*

$$E[f_i(\Delta)] \quad \geq \quad E[f_i(0)] \quad + \quad \alpha(\max_{\substack{(j_1,j_2) \in R_i \times R_i \\ \nexists k \in R_i : \Delta_{j_1} < \Delta_k < \Delta_{j_2}}} (\Delta_{j_2} - \Delta_{j_1})).$$

Proof Let $n := |R_i|$. By the linearity of the expected value, $E[f_i(\Delta)]$ can be rewritten in the following way:

$$E[f_i(\Delta)] = E\left[\sum_{j \in R_i}\left(d_{i,j} + \Delta_j - \min_{k \in R_i}(d_{i,k} + \Delta_k)\right)\right]$$

$$= E\left[\sum_{j \in R_i} d_{i,j}\right] + E\left[\sum_{j \in R_i} \Delta_j\right] - n \cdot E\left[\min_{j \in R_i}(d_{i,j} + \Delta_j)\right]$$

$$= \frac{n}{\lambda} + \sum_{j \in R_i} \Delta_j - n \cdot E\left[\min_{j \in R_i}(d_{i,j} + \Delta_j)\right].$$

Assume without loss of generality that $R_i = \{1, \ldots, n\}$ and that the nodes are ordered such that $\forall j, 1 \leq j < n : \Delta_{j+1} \geq \Delta_j$. The exponential distributions if the $d_{i,j}$ all share the same parameter λ, thus with Lemma B.10 we have

$$E[f_i(\Delta)] = \frac{n}{\lambda} + \sum_{j=1}^{n} \Delta_j - n\Delta_1$$

$$- n \cdot \sum_{j=1}^{n-1}\left(\prod_{k=1}^{j-1} e^{-k\lambda(\Delta_{k+1} - \Delta_k)}\right)\left(1 - e^{-j\lambda(\Delta_{j+1} - \Delta_j)}\right)\frac{1}{j\lambda}$$

$$- n \cdot \left(\prod_{k=1}^{n-1} e^{-k\lambda(\Delta_{k+1} - \Delta_k)}\right)\frac{1}{n\lambda}.$$

195

We now define $\forall j, 1 \leq j < n : \delta_j := \Delta_{j+1} - \Delta_j$ and $\delta := (\delta_1, \ldots, \delta_{n-1})^T$. Then $\forall j, 1 \leq j < n : \delta_j \geq 0$. (Note that δ has dimension zero for $n = 1$.) With the given definition of δ the following reformulation of $E[f_i(\Delta)]$ is possible:

$$E[f_i(\Delta)] = \frac{n}{\lambda} + \sum_{j=1}^{n-1} (n-j)\delta_j$$
$$- \sum_{j=1}^{n-1} \left(\prod_{k=1}^{j-1} e^{-k\lambda\delta_k} \right) \left(1 - e^{-j\lambda\delta_j} \right) \frac{n}{j\lambda} - \left(\prod_{k=1}^{n-1} e^{-k\lambda\delta_k} \right) \frac{1}{\lambda}.$$

We generalize this to a function K of arbitrary vectors θ with non-negative elements and dimension $m := \dim\theta$, i. e.

$$K(m,\theta) := \frac{m+1}{\lambda} + \sum_{j=1}^{m} (m-j+1)\theta_j$$
$$- \sum_{j=1}^{m} \left(\prod_{k=1}^{j-1} e^{-k\lambda\theta_k} \right) \left(1 - e^{-j\lambda\theta_j} \right) \frac{m+1}{j\lambda} - \left(\prod_{k=1}^{m} e^{-k\lambda\theta_k} \right) \frac{1}{\lambda}.$$

Note that

$$E[f_i(\Delta)] = K(n-1, \delta).$$

Now we construct α. Let e_k^d be the k-th unit vector of dimension d and set

$$\alpha(t) := \min_{\substack{1 \leq d < |J| \\ 1 \leq k \leq d}} \left(K(d, t \cdot e_k^d) - K(d, 0) \right).$$

For $n = 1$ the assertion is true since for $n = 1$

$$\max_{\substack{(j_1, j_2) \in R_i \times R_i \\ \nexists k \in R_i : \Delta_{j_1} < \Delta_k < \Delta_{j_2}}} (\Delta_{j_2} - \Delta_{j_1}) = 0$$

and $\alpha(0) = 0$.

We thus focus on the case $n > 1$ now. Note that in this case

$$\max_{\substack{(j_1, j_2) \in R_i \times R_i \\ \nexists k \in R_i : \Delta_{j_1} < \Delta_k < \Delta_{j_2}}} (\Delta_{j_2} - \Delta_{j_1}) = ||\delta||_\infty.$$

196

For each m, K is differentiable in δ. We will now show that for all $m \geq 1$ and all p with $1 \leq p \leq m$

$$\frac{\partial K(m, \theta)}{\partial \theta_p} > 0$$

for $\theta_p > 0$. Calculating the partial derivative explicitly yields

$$
\begin{aligned}
\frac{\partial K(\theta)}{\partial \theta_p} &= (m - p + 1) + p \sum_{j=p+1}^{m} \left(\prod_{k=1}^{j-1} e^{-k\lambda\theta_k} \right) \left(1 - e^{-j\lambda\theta_j} \right) \frac{m+1}{j} \\
&\quad - \left(\prod_{k=1}^{p} e^{-k\lambda\theta_k} \right) (m+1) + p \left(\prod_{k=1}^{m} e^{-k\lambda\theta_k} \right) \\
&= m - p + 1 \\
&\quad + p \left(\prod_{k=1}^{p} e^{-k\lambda\theta_k} \right) \left(\sum_{j=p+1}^{m} \left(\prod_{k=p+1}^{j-1} e^{-k\lambda\theta_k} \right) \left(1 - e^{-j\lambda\theta_j} \right) \frac{m+1}{j} \right. \\
&\quad \left. - \frac{m+1}{p} + \left(\prod_{k=p+1}^{m} e^{-k\lambda\theta_k} \right) \right).
\end{aligned}
$$

Since $\frac{m+1}{j} > 1$ for $1 \leq j \leq m$ and

$$\sum_{j=p+1}^{m} \left(\prod_{k=p+1}^{j-1} e^{-k\lambda\theta_k} \right) \left(1 - e^{-j\lambda\theta_j} \right) + \left(\prod_{k=p+1}^{m} e^{-k\lambda\theta_k} \right) = 1$$

the following holds:

$$
\begin{aligned}
\frac{\partial K(\theta)}{\partial \theta_p} &\geq m - p + 1 + p \underbrace{\left(\prod_{k=1}^{p} e^{-k\lambda\theta_k} \right)}_{\leq 1} \underbrace{\left(1 - \frac{m+1}{p} \right)}_{<0} \\
&\geq m - p + 1 + p \left(1 - \frac{m+1}{p} \right) \\
&= 0.
\end{aligned}
$$

Here, the first inequality is strict if there is a $j > p$ such that $\theta_j > 0$, and the second inequality is strict if there is a $j \leq p$ such that $\theta_j > 0$.

Now the inequality from the assertion is easily verified. Let r $(1 \leq r < n)$ be an index for which $\delta_r = ||\delta||_\infty$. Such an r exists by the definition of $|| \cdot ||_\infty$. Then

$$
E[f_i(\Delta)] - E[f_i(0)]
$$
$$
= \quad K(|R_i|, \delta) - K(|R_i|, 0)
$$
$$
\geq \quad K(|R_i|, ||\delta||_\infty \cdot e_r^{|R_i|}) - K(|R_i|, 0)
$$
$$
\geq \quad \min_{1 \leq k \leq |R_i|} \left(K(|R_i|, ||\delta||_\infty \cdot e_k^{|R_i|}) - K(|R_i|, 0) \right)
$$
$$
\geq \quad \alpha(||\delta||_\infty).
$$

The first inequality stems from the fact that all entries of the Jacobian of K are non-negative and δ is component-wise greater than $||\delta||_\infty \cdot e_r^{|R_i|}$.

It remains to show that $\alpha(||\delta||_\infty) > 0$ is strictly monotonically increasing. This, however, is easy to see. From the above calculations we know that the entries of the Jacobian of $K(m, \theta)$ are *strictly* positive if there is some k such that $\theta_k > 0$. $\alpha(t)$ is the minimum of

$$
K(d, t \cdot e_k^d) - K(d, 0)
$$

for a finite number of combinations of d and k. For each of these combinations and for any $t_1 > t_2 \geq 0$, though, $(K(d, t_1 \cdot e_k^d) - K(d, 0)) > (K(d, t_2 \cdot e_k^d) - K(d, 0))$, since the k-th component of $t_1 \cdot e_k^d$ is strictly greater than the respective component of $t_2 \cdot e_k^d$, whereas all other components are equal. Thus, $\alpha(t_1) > \alpha(t_2)$ and the assertion holds. $\square$

Lemma B.13. *There is some $\mathcal{L} \in \mathbb{R}$ such that*

$$
\frac{1}{|I|} \cdot k(L) = \frac{1}{|I|} \sum_{(i,j) \in R} \widehat{d}_{i,j}
$$

is Lipschitz continuous in Δ with Lipschitz constant $\mathcal{L}$. $\mathcal{L}$ does not depend on $|I|$.

Proof From the closed form of f_i given above it can be seen that f_i is continuous in Δ for each event $i \in I$. It is also easy to see that the partial derivatives of $f_i(\Delta)$ exist almost everywhere and are, where they exist, bounded above by 1 and below by $-|R_i|$, and thus also by $-|J|$. Thus, all f_i are Lipschitz continuous with a common Lipschitz constant $\mathcal{L}$.

Since
$$k(L) = \sum_{i \in I} f_i(\Delta)$$
we can conclude that $k(L)$ is Lipschitz continuous with Lipschitz constant $|I| \cdot \mathcal{L}$. Therefore, $\mathcal{L}$ is also a Lipschitz constant for $\frac{1}{|I|} \cdot k(L)$. $\qquad \square$

From the definition of f_i above it can be seen that $f_i(\Delta) = f_i(\Delta + (t, \dots, t)^T)$ for any $t \in \mathbb{R}$. This is again the offset ambiguity. Thus, from now on we can assume without loss of generality that the estimation error for the $|J|$-th node is zero. Therefore, we will ignore this node in the following and reduce Δ to a vector of dimension $|J| - 1$. Consistency of the MLE is then equivalent to

$$\operatorname*{plim}_{|I| \to \infty} ||\Delta|| = 0.$$

In the following theorem we will use infinity norm spheres. Our notation for them is as follows.

Definition

$$\forall m \in \mathbb{R}^n, r \in \mathbb{R} : S_\infty(m, r) := \left\{ x \mid ||x - m||_\infty < r \right\}$$
$$\forall m \in \mathbb{R}^n, r \in \mathbb{R} : \bar{S}_\infty(m, r) := \left\{ x \mid ||x - m||_\infty = r \right\}$$

Theorem B.14. *If the regularity condition is fulfilled, then for all $\epsilon, \delta > 0$ there exists an $N \in \mathbb{N}$ such that from $|I| \geq N$ it follows that for*

$$\tilde{\Delta} := \operatorname*{argmin}_{\Delta \in \mathbb{R}^j} \sum_{i \in I} f_i(\Delta)$$

the following holds

$$P(||\tilde{\Delta}||_\infty > \delta) < \epsilon$$

i. e.,

$$\operatorname*{plim}_{|I| \to \infty} ||\Delta|| = 0.$$

Thus, the simplified MLE is consistent.

Proof Let

$$r := \frac{\beta \cdot \alpha \left(\frac{\delta}{(|J|-1)^2} \right)}{3 \cdot \mathcal{L}}$$

with β from the regularity condition, α from Lemma B.12 and $\mathcal{L}$ from Lemma B.13.

Let $M \subset \bar{S}_\infty(0, \delta) \subset \mathbb{R}^{|J|-1}$ be a finite set of points such that

$$\bar{S}_\infty(0, \delta) \subset \bigcup_{m \in M} S_\infty(m, r).$$

Such a set of points exists since $\bar{S}_\infty(0, \delta)$ is compact and $r > 0$. Let m be one of the points in M. Then there exists $p \in J$ such that $m_p = \delta$. Let $G = (J, V)$ be the graph from the regularity condition. Then there is a path from p to the node with ID $|J|$ with a length of at most $|J| - 1$. The estimation error of the $|J|$-th node is 0 by assumption. Thus, there exists $\{j_1, j_2\} \in V$ such that $|m_{j_1} - m_{j_2}| \geq \frac{\delta}{|J|-1}$.

Let i be an event for which $\{j_1, j_2\} \subseteq R_i$. Since the total number of nodes is $|J|$, there are at most $|J|$ nodes in R_i, and for two of them, j_1 and j_2, it holds that $|m_{j_1} - m_{j_2}| \geq \frac{\delta}{|J|-1}$. Thus,

$$\max_{\substack{(j_1', j_2') \in R_i \times R_i \\ \nexists k \in R_i : m_{j_1'} < m_k < m_{j_2'}}} (m_{j_2'} - m_{j_1'}) \geq \frac{\delta}{(|J|-1)^2}.$$

We define

$$I^+ := \{i \in I | \{j_1, j_2\} \subseteq R_i\}.$$

From the regularity condition we know that

$$E[|I^+|] \geq \beta \cdot |I|.$$

This yields

$$E \left[\frac{1}{|I|} \sum_{i \in I} f_i(m) \right]$$

$$= \frac{1}{|I|} \left(E \left[\sum_{i \in I \setminus I^+} f_i(m) \right] + E \left[\sum_{i \in I^+} f_i(m) \right] \right)$$

$$\geq \frac{1}{|I|} \left(E \left[\sum_{i \in I \setminus I^+} f_i(0) \right] + E \left[\sum_{i \in I^+} \left(f_i(0) + \alpha \left(\frac{\delta}{(|J|-1)^2} \right) \right) \right] \right)$$

$$= \frac{1}{|I|} \left(E \left[\sum_{i \in I \setminus I^+} f_i(0) \right] + E \left[\sum_{i \in I^+} f_i(0) + |I^+| \cdot \alpha \left(\frac{\delta}{(|J| - 1)^2} \right) \right] \right)$$

$$= \frac{1}{|I|} \left(E \left[\sum_{i \in I^+} f_i(0) \right] + E \left[\sum_{i \in I \setminus I^+} f_i(0) \right] + E[|I^+|] \cdot \alpha \left(\frac{\delta}{(|J| - 1)^2} \right) \right)$$

$$\geq \frac{1}{|I|} \left(E \left[\sum_{i \in I} f_i(0) \right] + \beta \cdot |I| \cdot \alpha \left(\frac{\delta}{(|J| - 1)^2} \right) \right)$$

$$= E \left[\frac{1}{|I|} \sum_{i \in I} f_i(0) \right] + \beta \cdot \alpha \left(\frac{\delta}{(|J| - 1)^2} \right).$$

From the Lipschitz condition established in Lemma B.13 and the definition of r above we get that for all $m' \in \mathbb{R}^{|J|-1}$ with $||m' - m|| < r$ the following holds:

$$\left| \frac{1}{|I|} \sum_{i \in I} f_i(m) - \frac{1}{|I|} \sum_{i \in I} f_i(m') \right| < r \cdot \mathcal{L} = \frac{1}{3} \cdot \beta \cdot \alpha \left(\frac{\delta}{(|J| - 1)^2} \right).$$

Thus, we can conclude that if

$$\frac{1}{|I|} \sum_{i \in I} f_i(m) > E \left[\frac{1}{|I|} \sum_{i \in I} f_i(0) \right] + \frac{2}{3} \cdot \beta \cdot \alpha \left(\frac{\delta}{(|J| - 1)^2} \right) \tag{B.7}$$

holds for m, then for all $m' \in S_\infty(m, r)$

$$\frac{1}{|I|} \sum_{i \in I} f_i(m') > E \left[\frac{1}{|I|} \sum_{i \in I} f_i(0) \right] + \frac{1}{3} \cdot \beta \cdot \alpha \left(\frac{\delta}{(|J| - 1)^2} \right). \tag{B.8}$$

By the law of large numbers there is some $N_m \in \mathbb{N}$ for m where for $|I| \geq N_m$ it follows that

$$P \left(\frac{1}{|I|} \sum_{i \in I} f_i(m) > E \left[\frac{1}{|I|} \sum_{i \in I} f_i(0) \right] + \frac{2}{3} \cdot \beta \cdot \alpha \left(\frac{\delta}{(|J| - 1)^2} \right) \right) \geq 1 - \frac{\epsilon}{|M| + 1}.$$

We call the condition in the probability above the superiority condition for m. Similar to the superiority conditions, there is an inferiority condition: there exists some $N_0 \in \mathbb{N}$ such that for $|I| \geq N_0$

$$P \left(\frac{1}{|I|} \sum_{i \in I} f_i(0) < E \left[\frac{1}{|I|} \sum_{i \in I} f_i(0) \right] + \frac{1}{3} \cdot \beta \cdot \alpha \left(\frac{\delta}{(|J| - 1)^2} \right) \right) \geq 1 - \frac{\epsilon}{|M| + 1}.$$

Because $|M|$ is finite, there is some $N^* = \max(\{N_0\} \cup \{N_m | m \in M\})$ fulfilling the superiority conditions for all $m \in M$ as well as the inferiority condition, each with a probability of at least $1 - \frac{\epsilon}{|M|+1}$. Thus, for $|I| \geq N^*$ the probability that all $|M| + 1$ conditions are fulfilled is at least

$$1 - (|M| + 1) \cdot \frac{\epsilon}{|M| + 1} = 1 - \epsilon.$$

Since the spheres $S_\infty(m, r)$ around all $m \in M$ cover $\bar{S}_\infty(0, \delta)$, (B.8) holds for all $m' \in \bar{S}_\infty(0, \delta)$ if (B.7) holds for all $m \in M$. Hence, with probability of at least $1 - \epsilon$, it holds that

$$\forall m' \in \bar{S}_\infty(0, \delta) : \frac{1}{|I|} \sum_{i \in I} f_i(m') > \frac{1}{|I|} \sum_{i \in I} f_i(0).$$

Therefore, we know that $\frac{1}{|I|} \sum_{i \in I} f_i$ has a local optimum in $S_\infty(0, \delta)$. Since f_i is convex for each i by Lemma B.11, $\frac{1}{|I|} \sum_{i \in I} f_i$ is also convex. Thus, the local optimum is also a global optimum. A global optimum of $\frac{1}{|I|} \sum_{i \in I} f_i$ is also a global optimum of $\sum_{i \in I} f_i$. Therefore, for the vector $\tilde{\Delta}$ from the optimal LP solution we have

$$P(\tilde{\Delta} \in S_\infty(0, \delta)) = P(||\tilde{\Delta}|| \leq \delta) \geq 1 - \epsilon.$$

This is the assertion. $\qquad\qquad\qquad\qquad\qquad\qquad\qquad\qquad\qquad\qquad\qquad\qquad\quad \square$

From the consistency result regarding the clock offsets it is easy to obtain a result on the quality of the event time estimates in the same asymptotic scenario. It has been stated before that the estimates of the event times are given by $\widehat{T}_i = T_i + \min_{k \in R_i}(d_{i,k} + \Delta_k)$. Thus, if Δ is close to zero (neglecting the offset ambiguity), the estimate for event i will be wrong by $\min_{k \in R_i} d_{i,k}$. From Lemma B.7 it then follows that the error is exponentially distributed with parameter $|R_i| \cdot \lambda$. In particular this means that—as could be expected—the expected estimation error decreases with the number of nodes observing the same event.

Bibliography

Own Publications

[ASM]   Nasir Ali, Björn Scheuermann, and Martin Mauve. A witness system for vehicular ad hoc networks. Submitted for publication.

[CBSM07]   Murat Caliskan, Andreas Barthels, Björn Scheuermann, and Martin Mauve. Predicting parking lot occupancy in vehicular ad hoc networks. In *VTC '07-Spring: Proceedings of the 65th IEEE Vehicular Technology Conference*, pages 277–281, April 2007.

[JKM⁺]   Florian Jarre, Wolfgang Kiess, Martin Mauve, Magnus Roos, and Björn Scheuermann. Least squares timestamp synchronization for local broadcast networks. Submitted for publication.

[JSLM06a]   Yves Igor Jerschow, Björn Scheuermann, Christian Lochert, and Martin Mauve. A cross-layer protocol evaluation framework on ESB nodes (demo). In *REALMAN '06: Proceedings of the 2nd International Workshop on Multi-hop Ad Hoc Networks: from Theory to Reality*, pages 104–106, May 2006.

[JSLM06b]   Yves Igor Jerschow, Björn Scheuermann, Christian Lochert, and Martin Mauve. A real-world framework to evaluate cross-layer protocols for wireless multihop networks. In *REALMAN '06: Proceedings of the 2nd International Workshop on Multi-hop Ad Hoc Networks: from Theory to Reality*, pages 1–6, May 2006.

[LSC⁺05]   Christian Lochert, Björn Scheuermann, Murat Caliskan, Andreas Barthels, Alfonso Cervantes, and Martin Mauve. Multiple simulator interlinking environment for IVC. In *VANET '05: Proceedings of the 2nd ACM International Workshop on Vehicular Ad Hoc Networks*, pages 87–88, September 2005.

[LSCM07]   Christian Lochert, Björn Scheuermann, Murat Caliskan, and Martin Mauve. The feasibility of information dissemination in vehicular ad-hoc networks. In *WONS '07: Proceedings of the 4th Annual Conference on Wireless On-demand Network Systems and Services*, pages 92–99, January 2007.

[LSM07a] Christian Lochert, Björn Scheuermann, and Martin Mauve. Probabilistic aggregation for data dissemination in VANETs. In *VANET '07: Proceedings of the 4th ACM International Workshop on Vehicular Ad Hoc Networks*, pages 1–8, September 2007.

[LSM07b] Christian Lochert, Björn Scheuermann, and Martin Mauve. A survey on congestion control for mobile ad-hoc networks. *Wiley Wireless Communications and Mobile Computing*, 7(5):655–676, June 2007.

[LSSM07] Peter Lieven, Björn Scheuermann, Michael Stini, and Martin Mauve. Filtering spam email based on retry patterns. In *ICC '07: Proceedings of the IEEE International Conference on Communications*, pages 1515–1520, June 2007.

[RSK+07] Jedrzej Rybicki, Björn Scheuermann, Wolfgang Kiess, Christian Lochert, Pezhman Fallahi, and Martin Mauve. Challenge: Peers on wheels—a road to new traffic information systems. In *MobiCom '07: Proceedings of the 13th Annual ACM International Conference on Mobile Computing and Networking*, pages 215–221, September 2007.

[Sch07] Björn Scheuermann. *Visualisierung von MANET-Simulationen – Analyse von ns-2-Tracefiles mit Huginn*. VDM Verlag Dr. Müller, Saarbrücken, Germany, 2007. In German language.

[SFT+05a] Björn Scheuermann, Holger Füßler, Matthias Transier, Marcel Busse, Martin Mauve, and Wolfgang Effelsberg. Huginn: A 3D visualizer for wireless ns-2 traces. In *MSWiM '05: Proceedings of the 8th ACM International Symposium on Modeling, Analysis and Simulation of Wireless and Mobile Systems*, pages 143–150, October 2005.

[SFT+05b] Björn Scheuermann, Holger Füßler, Matthias Transier, Martin Mauve, and Wolfgang Effelsberg. Visualizing wireless ns-2 traces in 3D. In *MobiCom '05: The 11th Annual ACM International Conference on Mobile Computing and Networking, Demo Session*, September 2005.

[SHC07] Björn Scheuermann, Wenjun Hu, and Jon Crowcroft. Near-optimal coordinated network coding in wireless multihop networks. In *CoNEXT '07: Proceedings of the 3rd International Conference on Emerging Networking Experiments and Technologies*, December 2007.

[SKLM] Björn Scheuermann, Markus Koegel, Christian Lochert, and Martin Mauve. Reliable wireless multihop communication without end-to-end acknowledgments. Submitted for publication.

[SKR+a] Björn Scheuermann, Wolfgang Kiess, Magnus Roos, Florian Jarre, and Martin Mauve. Error bounds and consistency in maximum likelihood time synchronization. Technical report, Department of Computer Science, Heinrich Heine University Düsseldorf, Germany. To appear.

204

[SKR⁺b]   Björn Scheuermann, Wolfgang Kiess, Magnus Roos, Florian Jarre, and
            Martin Mauve.   On the time synchronization of distributed logfiles in
            networks with local broadcast media. Submitted for publication.

[SLM08]   Björn Scheuermann, Christian Lochert, and Martin Mauve. Implicit hop-
            by-hop congestion control in wireless multihop networks. *Elsevier Ad Hoc
            Networks*, 6(2):260–286, April 2008.

[SM07]   Björn Scheuermann and Martin Mauve.   Near-optimal compression of
            Flajolet-Martin sketches. In *Dial M-POMC '07: Proceedings of the 4th
            ACM SIGACT-SIGOPS International Workshop on Foundations of Mo-
            bile Computing*, August 2007.

[STL⁺07]   Björn Scheuermann, Matthias Transier, Christian Lochert, Martin Mauve,
            and Wolfgang Effelsberg.   Backpressure multicast congestion control in
            mobile ad-hoc networks. In *CoNEXT '07: Proceedings of the 3rd Interna-
            tional Conference on Emerging Networking Experiments and Technologies*,
            December 2007.

[TSM07]   Thi Minh Chau Tran, Björn Scheuermann, and Martin Mauve. Detecting
            the presence of nodes in MANETs.   In *CHANTS '07: Proceedings of
            the 3rd ACM MobiCom Workshop on Challenged Networks*, pages 43–50,
            September 2007.

Other References

[AACP05]   Guiseppe Anastasi, Emilio Ancillotti, Marco Conti, and Andrea Passarella.
            TPA: A transport protocol for ad hoc networks. In *ISCC '05: Proceedings
            of the 10th IEEE International Symposium on Computers and Communi-
            cation*, pages 51–56, June 2005.

[ACDM]   David Applegate, William Cook, Sanjeeb Dash, and Monika Mevenkamp.
            QSopt linear programming solver.   Version 1.01.   http://www2.isye.-
            gatech.edu/~wcook/qsopt/.

[ACLY00]   Rudolf Ahlswede, Ning Cai, Shuo-Yen Li, and Raymond Yeung. Network
            information flow. *IEEE Transactions on Information Theory*, 46(4):1204–
            1216, 2000.

[AJ03]   Eitan Altman and Tania Jiménez.   Novel delayed ACK techniques for
            improving TCP performance in multihop wireless networks. In *PWC '03:
            Proceedings of the IFIP-TC6 8th International Conference on Personal
            Wireless Communications*, pages 237–250, September 2003.

[APS99]   Mark Allman, Vern Paxson, and W. Richard Stevens. TCP congestion control. RFC 2581 (Proposed Standard), April 1999. Updated by RFC 3390.

[APSS04]   Vaidyanathan Anantharaman, Seung-Jong Park, Karthikeyan Sundaresan, and Raghupathy Sivakumar. TCP performance over mobile ad-hoc networks: a quantitative study. *Wiley Wireless Communications and Mobile Computing*, 4(2):203–222, 2004.

[Ash95]   Paul Ashton. Algorithms for off-line clock synchronization. Technical Report TR COSC 12/95, Department of Computer Science, University of Canterbury, December 1995.

[Bau05]   Peter Baumung. Stable, congestion-controlled application-layer multicasting in pedestrian ad-hoc networks. In *WoWMoM '05: Proceedings of the 6th IEEE International Symposium on a World of Wireless Mobile and Multimedia Networks*, pages 57–64, June 2005.

[Bra89]   Robert Braden. Requirements for Internet hosts – communication layers. RFC 1122 (Standard), October 1989. Updated by RFCs 1349, 4379.

[BV05]   Saâd Biaz and Nitin H. Vaidya. "De-randomizing" congestion losses to improve TCP performance over wired-wireless networks. *IEEE/ACM Transactions on Networking*, 13(3):596–608, June 2005.

[BYS+04]   Dan Berger, Zhenqiang Ye, Prasun Sinha, Srikanth Krishnamurthy, Michaelis Faloutsos, and Satish K. Tripathi. TCP-friendly medium access control for ad-hoc wireless networks: Alleviating self-contention. In *MASS '04: Proceedings of the 1st International Conference on Mobile Ad hoc and Sensor Systems*, pages 214–223, October 2004.

[BZK04]   Peter Baumung, Martina Zitterbart, and Kendy Kutzner. Improving delivery ratios for application layer multicast in mobile ad-hoc networks. In *ASWN '04: Proceedings of the 4th Workshop on Applications and Services in Wireless Networks*, pages 132–141, August 2004.

[CN02]   Kai Chen and Klara Nahrstedt. EXACT: An explicit rate-based flow control framework in MANET (extended version). Technical Report UIUCDCS-R-2002-2286/UILU-ENG-2002-1730, Department of Computer Science, University of Illinois at Urbana-Champaign, July 2002.

[CN04]   Kai Chen and Klara Nahrstedt. Limitations of equation-based congestion control in mobile ad hoc networks. In *ICDCSW '04: Proceedings of the 24th International Conference on Distributed Computing Systems Workshops - W7: EC*, pages 756–761, 2004.

[CNV04]   Kai Chen, Klara Nahrstedt, and Nitin H. Vaidya. The utility of explicit rate-based flow control in mobile ad hoc networks. In *WCNC '04: Proceedings of the IEEE Wireless Communications and Networking Conference*, volume 3, pages 1921–1926, March 2004.

[CP07]   Prasanna Chaporkar and Alexandre Proutiere. Adaptive network coding
             and scheduling for maximizing throughput in wireless networks. In *Mobi-
             Com '07: Proceedings of the 13th Annual ACM International Conference
             on Mobile Computing and Networking*, pages 135–146, September 2007.

[CRVP98]   Kartik Chandran, Sudarshan Raghunathan, Subbarayan Venkatesan, and
             Ravi Prakash. A feedback based scheme for improving TCP performance
             in ad-hoc wireless networks. In *ICDCS '98: Proceedings of the 18th In-
             ternational Conference on Distributed Computing Systems*, pages 472–479,
             May 1998.

[CXN03]   Kai Chen, Yuan Xue, and Klara Nahrstedt. On setting TCP's congestion
             window limit in mobile ad hoc networks. In *ICC '03: Proceedings of
             the IEEE International Conference on Communications*, volume 2, pages
             1080–1084, May 2003.

[DAVR04]   Adam Dunkels, Juan Alonso, Thiemo Voigt, and Hartmut Ritter. Dis-
             tributed TCP caching for wireless sensor networks. In *MedHocNet '04:
             Proceedings of the 3rd Annual Mediterranean Ad Hoc Networking Work-
             shop*, June 2004.

[DB01]   Thomas D. Dyer and Rajendra V. Boppana. A comparison of TCP per-
             formance over three routing protocols for mobile ad hoc networks. In
             *MobiHoc '01: Proceedings of the 2nd ACM International Symposium on
             Mobile Ad Hoc Networking and Computing*, pages 56–66, June 2001.

[DB04]   Saman Desilva and Rajendra V. Boppana. On the impact of noise sen-
             sitivity on performance in 802.11 based ad hoc networks. In *ICC '04:
             Proceedings of the IEEE International Conference on Communications*,
             volume 7, pages 4372–4376, June 2004.

[DHHB87]   Andrzej Duda, Gilbert Harrus, Yoram Haddad, and Guy Bernard. Esti-
             mating global time in distributed systems. In *ICDCS '87: Proceedings of
             the 7th International Conference on Distributed Computing Systems*, pages
             299–306, September 1987.

[dMCDA02]   Carlos de M. Cordeiro, Samir R. Das, and Dharma P. Agrawal. COPAS:
             Dynamic contention-balancing to enhance the performance of TCP over
             multi-hop wireless networks. In *ICCCN '02: Proceedings of the 11th Inter-
             national Conference on Computer Communications and Networks*, pages
             382–387, October 2002.

[dOB02]   Ruy de Oliveira and Torsten Braun. TCP in wireless mobile ad hoc
             networks. Technical Report IAM-02-003, Institute of Computer Science
             and Applied Mathematics, University of Berne, July 2002.

[dOB04]   Ruy de Oliveira and Torsten Braun. A delay-based approach using fuzzy
             logic to improve TCP error detection in ad hoc networks. In *WCNC '04:*

Proceedings of the IEEE Wireless Communications and Networking Conference, volume 3, pages 1666–1671, March 2004.

[dOB07]   Ruy de Oliveira and Torsten Braun. A smart TCP acknowledgment approach for multihop wireless networks. *IEEE Transactions on Mobile Computing*, 6(2):192–205, February 2007.

[dOBH03]   Ruy de Oliveira, Torsten Braun, and Marc Heissenbüttel. An edge-based approach for improving TCP in wireless mobile ad hoc networks. In *DADS '03: Proceedings of the Conference on Design, Analysis and Simulation of Distributed Systems*, March 2003.

[EGE02]   Jeremy Elson, Lewis Girod, and Deborah Estrin. Fine-grained network time synchronization using reference broadcasts. In *OSDI '02: Fifth USENIX Symposium on Operating Systems Design and Implementation*, pages 147–163, December 2002.

[EKL05]   Sherif M. ElRakabawy, Alexander Klemm, and Christoph Lindemann. TCP with adaptive pacing for multihop wireless networks. In *MobiHoc '05: Proceedings of the 6th ACM International Symposium on Mobile Ad Hoc Networking and Computing*, pages 288–299, May 2005.

[FGML02]   Zhenghua Fu, Benjamin Greenstein, Xiaoqiao Meng, and Songwu Lu. Design and implementation of a TCP-friendly transport protocol for ad hoc wireless networks. In *ICNP '02: Proceedings of the 10th IEEE International Conference on Network Protocols*, pages 216–225, November 2002.

[FHG04]   Sally Floyd, Tom Henderson, and Andrei Gurtov. The newreno modification to TCP's fast recovery algorithm. RFC 3782 (Proposed Standard), April 2004.

[FHPW00]   Sally Floyd, Mark Handley, Jitendra Padhye, and Jörg Widmer. Equation-based congestion control for unicast applications. In *SIGCOMM '00: Proceedings of the 2000 Conference on Applications, Technologies, Architectures, and Protocols for Computer Communications*, pages 43–56, August 2000.

[FJ93]   Sally Floyd and Van Jacobson. Random early detection gateways for congestion avoidance. *IEEE/ACM Transactions on Networking*, 1(4):397–413, August 1993.

[FML02]   Zhenghua Fu, Xiaoqiao Meng, and Songwu Lu. How bad TCP can perform in mobile ad hoc networks. In *ISCC '02: Proceedings of the 7th IEEE International Symposium on Computers and Communication*, pages 298–303, July 2002.

[FML03]   Zhenghua Fu, Xiaoqiao Meng, and Songwu Lu. A transport protocol for supporting multimedia streaming in mobile ad-hoc networks. *IEEE Journal on Selected Areas in Communications*, 21(10):1615–1626, December 2003.

[Fre]   Freie Universität Berlin, Computer Systems Telematics.   ScatterWeb
             project. http://www.inf.fu-berlin.de/inst/ag-tech/scatterweb_net.

[FZL⁺03]   Zhenghua Fu, Petros Zerfos, Haiyun Luo, Songwu Lu, Lixia Zhang, and
             Mario Gerla. The impact of multihop wireless channel on TCP throughput
             and loss. In *INFOCOM '03: Proceedings of the 22nd Annual Joint Con-
             ference of the IEEE Computer and Communications Societies*, volume 3,
             pages 1744–1753, March 2003.

[GAGPK03]   Tom Goff, Nael B. Abu-Ghazaleh, Dhananjay S. Phatak, and Ridvan
             Kahvecioglu. Preemptive routing in ad hoc networks. *Elsevier Paral-
             lel and Distributed Computing*, 63(2):123–140, February 2003.

[Git76]   Israel Gitman. Comparison of hop-by-hop and end-to-end acknowledgment
             schemes in computer communication networks. *IEEE Transactions on
             Communications*, 24(11):1258–1262, November 1976.

[GJ82]   Neil Gower and John Jubin. Congestion control using pacing in a packet
             radio network. In *MILCOM '82: Proceedings of the IEEE Military Com-
             munications Conference 'Progress in Spread Spectrum Communications'*,
             pages 23.1.1–23.1.6, October 1982.

[GNAA04]   Hrishikesh Gossain, Nagesh Nandiraju, Kumar Anand, and Dharma P.
             Agrawal.   Supporting MAC layer multicast in IEEE 802.11 based
             MANETs: Issues and solutions. In *LCN '04: Proceedings of the 29th An-
             nual IEEE International Conference on Local Computer Networks*, pages
             172–179, November 2004.

[GTB99]   Mario Gerla, Ken Tang, and Rajive Bagrodia. TCP performance in wire-
             less multi-hop networks. In *WMCSA '99: Proceedings of the 2nd IEEE
             Workshop on Mobile Computing Systems and Applications*, page 41, Febru-
             ary 1999.

[GV02]   Mesut Güneş and Donald Vlahovic. The performance of the TCP/RCWE
             enhancement for ad-hoc networks. In *ISCC '02: Proceedings of the 7th
             IEEE International Symposium on Computers and Communication*, pages
             43–48, July 2002.

[HCH06]   Tracey Ho, Yu-Han Chang, and Keesook Han. On constructive network
             coding for multiple unicasts. In *Proceedings of 44th Allerton Conference
             on Communication, Control and Computing*, 2006.

[HJB04]   Bret Hull, Kyle Jamieson, and Hari Balakrishnan. Mitigating conges-
             tion in wireless sensor networks. In *SenSys '04: Proceedings of the 2nd
             International Conference on Embedded Networked Sensor Systems*, pages
             134–147, November 2004.

[HSC95]   Hugh W. Holbrook, Sandeep K. Singhal, and David R. Cheriton. Log-based receiver-reliable multicast for distributed interactive simulation. In *SIGCOMM '95: Proceedings of the 1995 Conference on Applications, Technologies, Architectures, and Protocols for Computer Communications*, pages 328–341, August 1995.

[HUL91]   Jean-Baptiste Hiriart-Urruty and Claude Lemaréchal. *Convex Analysis and Minimization Algorithms I*. Springer, Berlin–Heidelberg–New York, 1991.

[HV99]   Gavin Holland and Nitin H. Vaidya. Analysis of TCP performance over mobile ad hoc networks. In *MobiCom '99: Proceedings of the 5th Annual ACM International Conference on Mobile Computing and Networking*, pages 219–230, July 1999.

[IAC99]   Sami Iren, Paul D. Amer, and Phillip T. Conrad. The transport layer: Tutorial and survey. *ACM Computing Surveys*, 31(4):360–405, December 1999.

[Jac88]   Van Jacobson. Congestion avoidance and control. In *SIGCOMM '88: Proceedings of the 1988 Conference on Applications, Technologies, Architectures, and Protocols for Computer Communications*, pages 314–329, August 1988.

[JCH84]   Rajendra K. Jain, Dah-Ming W. Chiu, and William R. Hawe. A quantitative measure and discrimination for resource allocation in shared computer systems. Technical Report DEC-TR-301, Digital Equipment Corporation, Eastern Research Lab, September 1984.

[JD06]   Shweta Jain and Samir R. Das. MAC layer multicast in wireless multi-hop networks. In *COMSWARE '06: Proceedings of the 1st International Conference on Communication System Software and Middleware*, January 2006.

[JM96]   David B. Johnson and David A. Maltz. Dynamic source routing in ad hoc wireless networks. In Thomasz Imielinski and Hank Korth, editors, *Mobile Computing*, volume 353, chapter 5, pages 153–181. Kluwer Academic Publishers, Norwell, MA, USA, 1996.

[KEPS04]   Richard M. Karp, Jeremy Elson, Christos H. Papadimitriou, and Scott Shenker. Global synchronization in sensornets. In *LATIN '04: Proceedings of the 6th Latin American Symposium on Theoretical Informatics*, pages 609–624, April 2004.

[KG06]   Dzmitry Kliazovich and Fabrizio Granelli. Cross-layer congestion control in ad hoc wireless networks. *Elsevier Ad Hoc Networks*, 4(6):687–708, November 2006.

[KHR02]   Dina Katabi, Mark Handley, and Charlie Rohrs.   Congestion control for
            high bandwidth-delay product networks. In *SIGCOMM '02: Proceedings
            of the 2002 Conference on Applications, Technologies, Architectures, and
            Protocols for Computer Communications*, pages 89–102, August 2002.

[KKFT02]   Swastik Kopparty, Srikanth V. Krishnamurthy, Michaelis Faloutsos, and
            Satish K. Tripathi. Split-TCP for mobile ad hoc networks. In *GLOBE-
            COM '02: Proceedings of the IEEE Global Telecommunications Confer-
            ence*, volume 1, pages 138–142, November 2002.

[KM07]   Wolfgang Kiess and Martin Mauve. Real-world evaluation of mobile ad-
            hoc networks.   In Marco Conti, Jon Crowcroft, and Andrea Passarella,
            editors, *Multi-hop Ad hoc Networks from Theory to Reality*, pages 1–22.
            Nova Science Publishers, Hauppauge, NY, USA, 2007.

[KO87]   Hermann Kopetz and Wilhelm Ochsenreiter.   Clock synchronization in
            distributed real-time systems. *IEEE Transactions on Communications*,
            36(8):933–940, 1987.

[KRH$^+$06] Sachin Katti, Hariharan Rahul, Wenjun Hu, Dina Katabi, Muriel Medard,
            and Jon Crowcroft. XORs in the air: Practical wireless network coding.
            In *SIGCOMM '06: Proceedings of the 2006 Conference on Applications,
            Technologies, Architectures, and Protocols for Computer Communications*,
            pages 243–254, September 2006.

[KTC00]   Dongkyun Kim, Chai-Keong Toh, and Yanghee Choi. TCP-BuS: Improv-
            ing TCP performance in wireless ad-hoc networks. In *ICC '00: Proceed-
            ings of the IEEE International Conference on Communications*, volume 3,
            pages 1707–1713, June 2000.

[KWC$^+$05] Christian Kreibich, Andrew Warfield, Jon Crowcroft, Steven Hand, and
            Ian Pratt. Using packet symmetry to curtail malicious traffic. In *Hot-
            Nets '05: Proceedings of the 4th Workshop on Hot Topics in Networks*,
            November 2005.

[KYKT05]   Fabius Klemm, Zhenqiang Ye, Srikanth V. Krishnamurthy, and Satish K.
            Tripathi. Improving TCP performance in ad hoc networks using signal
            strength based link management. *Elsevier Ad Hoc Networks*, 3(2):175–191,
            March 2005.

[LGC99]   Sung-Ju Lee, Mario Gerla, and Ching-Chuan Chiang. On-demand multi-
            cast routing protocol. In *WCNC '99: Proceedings of the IEEE Wireless
            Communications and Networking Conference*, volume 3, pages 1298–1302,
            September 1999.

[LL04]   Zongpeng Li and Baochun Li.   Network coding: The case for multiple
            unicast sessions. In *Allerton '04: Proceedings of the 42nd Annual Allerton
            Conference*, September 2004.

[LLA⁺04]   Mingzhe Li, Choong-Soo Lee, Emmanuel Agu, Mark Claypool, and Robert Kinicki. Performance enhancement of TFRC in wireless ad hoc networks. In *DMS '04: Proceedings of the 10th International Conference on Distributed Multimedia Systems*, September 2004.

[LS99]   Jian Liu and Suresh Singh. ATP: Application controlled transport protocol for mobile ad hoc networks. In *WCNC '99: Proceedings of the IEEE Wireless Communications and Networking Conference*, volume 3, pages 1318–1322, September 1999.

[LS01]   Jian Liu and Suresh Singh. ATCP: TCP for mobile ad hoc networks. *IEEE Journal on Selected Areas in Communications*, 19(7):1300–1315, July 2001.

[LXG03]   Haejung Lim, Kaixin Xu, and Mario Gerla. TCP performance over multipath routing in mobile ad hoc networks. In *ICC '03: Proceedings of the IEEE International Conference on Communications*, volume 2, pages 1064–1068, May 2003.

[Meh92]   Sanjay Mehrotra. On the implementation of a primal-dual interior point method. *SIAM Journal on Optimization*, 2(4):575–601, 1992.

[MFNT00]   Michael Mock, Reiner Frings, Edgar Nett, and Spiro Trikaliotis. Continuous clock synchronization in wireless real-time applications. In *SRDS '00: Proceedings of the 19th IEEE Symposium on Reliable Distributed Systems*, pages 125–132, October 2000.

[Mil92]   David L. Mills. Network time protocol (version 3) specification, implementation and analysis. RFC 1305 (Draft Standard), March 1992.

[Mil94]   David L. Mills. Internet time synchronization: The network time protocol. In Zhonghua Yang and T. Anthony Marsland, editors, *Global States and Time in Distributed Systems*. IEEE Computer Society Press, 1994.

[MSB00]   Jeffrey P. Monks, Prasun Sinha, and Vaduvur Bharghavan. Limitations of TCP-ELFN for ad hoc networks. In *MoMuC '00: Proceedings of the 7th IEEE International Workshop on Mobile Multimedia Communications*, October 2000.

[MST99]   Sue B. Moon, Paul Skelly, and Donald F. Towsley. Estimation and removal of clock skew from network delay measurements. In *INFOCOM '99: Proceedings of the 18th Annual Joint Conference of the IEEE Computer and Communications Societies*, pages 227–234, March 1999.

[MWH01]   Martin Mauve, Jörg Widmer, and Hannes Hartenstein. A survey on position-based routing in mobile ad-hoc networks. *IEEE Network*, 15(6):30–39, November 2001.

[NC04]   William Navidi and Tracy Camp. Stationary distributions for the random waypoint mobility model. *IEEE Transactions on Mobile Computing*, 3(1):99–108, January 2004.

[NHK05]   Kitae Nahm, Ahmed Helmy, and C.-C. Jay Kuo. TCP over multihop
            802.11 networks: Issues and performance enhancement. In *MobiHoc '05:
            Proceedings of the 6th ACM International Symposium on Mobile Ad Hoc
            Networking and Computing*, pages 277–287, May 2005.

[ns2a]   The ns-2 network simulator. http://www.isi.edu/nsnam/ns. version 2.30.

[ns2b]   Wireless multicast extensions for ns-2.1b8. http://www.monarch.cs.rice.-
            edu/multicast_extensions.html.

[NW99]   Jorge Nocedal and Stephen J. Wright. *Numerical Optimization*. Springer,
            Berlin, 1999.

[PAM+05]   Thierry Plesse, Cedric Adjih, Pascale Minet, Anis Laouiti, Adokoé Plakoo,
            Marc Badel, Paul Mühlethaler, Philippe Jacquet, and Jérôme Lecomte.
            OLSR performance measurement in a military mobile ad hoc network.
            Elsevier Ad Hoc Networks, 3(5):575–588, September 2005.

[PBRD03]   Charles E. Perkins, Elizabeth M. Belding-Royer, and Samir R. Das. Ad
            hoc on-demand distance vector (AODV) routing. RFC 3561 (Experimen-
            tal), July 2003.

[POK04]   Stylianos Papanastasiou and Mohamed Ould-Khaoua. TCP congestion
            window evolution and spatial reuse in MANETs. *Wiley Wireless Commu-
            nications and Mobile Computing*, 4(6):669–682, September 2004.

[PPW+07]   Manoj Pandey, Roger Pack, Lei Wang, Quiyi Duan, and Daniel Zappala.
            To repair or not to repair: Helping routing protocols to distinguish mobility
            from congestion. In *INFOCOM '07 Mini Symposia*, May 2007.

[PR99]   Charles E. Perkins and Elizabeth M. Royer. Ad-hoc on-demand distance
            vector routing. In *WMCSA '99: Proceedings of the 2nd IEEE Workshop
            on Mobile Computing Systems and Applications*, pages 90–100, February
            1999.

[PS03]   Jun Peng and Biplab Sikdar. A multicast congestion control scheme
            for mobile ad-hoc networks. In *GLOBECOM '03: Proceedings of the
            IEEE Global Telecommunications Conference*, volume 5, pages 2860–2864,
            December 2003.

[RBM05]   Kay Römer, Philipp Blum, and Lennart Meier. Time synchronization
            and calibration in wireless sensor networks. In Ivan Stojmenovic, editor,
            Handbook of Sensor Networks: Algorithms and Architectures, pages 199–
            237. John Wiley & Sons, September 2005.

[RK06]   Vivek Raghunathan and Panganamala R. Kumar. A counterexample in
            congestion control of wireless networks. *Elsevier Performance Evaluation*,
            64:399–418, June 2006.

[RL02]   Božidar Radunović and Jean-Yves Le Boudec. A unified framework for
             max-min and min-max fairness with applications. In *Allerton '02: Pro-
             ceedings of the 40th Annual Allerton Conference*, October 2002.

[ROY+04]   Venkatesh Rajendran, Katia Obraczka, Yunjung Yi, Sung-Ju Lee, Ken
             Tang, and Mario Gerla.   Combining source- and localized recovery to
             achieve reliable multicast in multi-hop ad hoc networks. In *NETWORK-
             ING '04: Proceedings of the 3rd International IFIP-TC6 Networking Con-
             ference*, pages 112–124, May 2004.

[RS07]   Saikat Ray and David Starobinski. On false blocking in RTS/CTS-based
             multi-hop wireless networks. *IEEE Transactions on Vehicular Technology*,
             56(2):849–862, March 2007.

[SAHS03]   Karthikeyan Sundaresan, Vaidyanathan Anantharaman, Hung-Yun Hsieh,
             and Raghupathy Sivakumar. ATP: A reliable transport protocol for ad-
             hoc networks. In *MobiHoc '03: Proceedings of the 4th ACM International
             Symposium on Mobile Ad Hoc Networking and Computing*, pages 64–75,
             June 2003.

[SG05]   Yang Su and Thomas Gross.   WXCP: Explicit congestion control for
             wireless multi-hop networks. In *IWQoS '05: Proceedings of the 12th In-
             ternational Workshop on Quality of Service*, June 2005.

[SM01]   Dong Sun and Hong Man. ENIC – an improved reliable transport scheme
             for mobile ad hoc networks. In *GLOBECOM '01: Proceedings of the
             IEEE Global Telecommunications Conference*, volume 5, pages 2852–2856,
             November 2001.

[SM03]   Masashi Sugano and Masayuki Murata. Performance improvement of TCP
             on a wireless ad hoc network. In *VTC '03-Spring: Proceedings of the 57th
             IEEE Vehicular Technology Conference*, volume 4, pages 2276–2280, April
             2003.

[SRB07]   Sudipta Sengupta, Shravan Rayanchu, and Suman Banerjee. An analy-
             sis of wireless network coding for unicast sessions: The case for coding-
             aware routing. In *INFOCOM '07: Proceedings of the 26th Annual Joint
             Conference of the IEEE Computer and Communications Societies*, pages
             1028–1036, May 2007.

[SRP]   Jos F. Sturm, Oleksandr Romanko, and Imre Pólik. SeDuMi. Version
             1.1R2. http://sedumi.mcmaster.ca/.

[TFW+07]   Matthias Transier, Holger Füßler, Jörg Widmer, Martin Mauve, and Wolf-
             gang Effelsberg. A hierarchical approach to position-based multicast for
             mobile ad-hoc networks. *Springer Wireless Networks*, 13(4), August 2007.

[TG03]   Ken Tang and Mario Gerla. Congestion control multicast in wireless ad hoc
             networks. *Elsevier Computer Communications*, 26(3):278–288, February
             2003.

[TOLG02a] Ken Tang, Katia Obraczka, Sung-Ju Lee, and Mario Gerla. Congestion controlled adaptive lightweight multicast in wireless mobile ad hoc networks. In *ISCC '02: Proceedings of the 7th IEEE International Symposium on Computers and Communication*, pages 967–972, July 2002.

[TOLG02b] Ken Tang, Katia Obraczka, Sung-Ju Lee, and Mario Gerla. A reliable, congestion-controlled multicast transport protocol in multimedia multihop networks. In *WPMC '02: Proceedings of the 5th International Symposium on Wireless Personal Multimedia Communications*, October 2002.

[TOLG03] Ken Tang, Katia Obraczka, Sung-Ju Lee, and Mario Gerla. Reliable adaptive lightweight multicast protocol. In *ICC '03: Proceedings of the IEEE International Conference on Communications*, pages 1054–1058, May 2003.

[VBP04] Darryl Veitch, Satish Babu, and Attila Pàsztor. Robust synchronization of software clocks across the Internet. In *IMC '04: Proceedings of the 4th ACM SIGCOMM Conference on Internet Measurement*, pages 219–232, October 2004.

[VRC97] Paulo Veríssimo, Luís Rodrigues, and Antonio Casimiro. Cesiumspray: a precise and accurate global time service for large-scale systems. *Real-Time Systems*, 12(3):243–294, 1997.

[WC01] Alec Woo and David E. Culler. A transmission control scheme for media access in sensor networks. In *MobiCom '01: Proceedings of the 7th Annual ACM International Conference on Mobile Computing and Networking*, pages 221–235, July 2001.

[WCK05] Yunnan Wu, Philip A. Chou, and Sun-Yuan Kung. Information exchange in wireless networks with network coding and physical-layer broadcasts. In *CISS '05: Proceedings of the 39th Annual Conference on Information Sciences and Systems*, March 2005.

[WEC03] Chieh-Yih Wan, Shane B. Eisenman, and Andrew T. Campbell. CODA: Congestion detection and avoidance in sensor networks. In *SenSys '03: Proceedings of the 1st International Conference on Embedded Networked Sensor Systems*, pages 266–279, November 2003.

[WSL+06] Chonggang Wang, Kazem Sohraby, Bo Li, Mahmoud Daneshmand, and Yueming Hu. A survey of transport protocols for wireless sensor networks. *IEEE Network*, 20(3):34–40, May 2006.

[WZ02] Feng Wang and Yongguang Zhang. Improving TCP performance over mobile ad-hoc networks with out-of-order detection and response. In *MobiHoc '02: Proceedings of the 3rd ACM International Symposium on Mobile Ad Hoc Networking and Computing*, pages 217–225, June 2002.

[XGB02]   Kaixin Xu, Mario Gerla, and Sang Bae. How effective is the IEEE 802.11
             RTS/CTS handshake in ad hoc networks? In *GLOBECOM '02: Pro-
             ceedings of the IEEE Global Telecommunications Conference*, pages 72–76,
             November 2002.

[XGQS05]   Kaixin Xu, Mario Gerla, Lantao Qi, and Yantai Shu. TCP unfairness
             in ad hoc wireless networks and a neighborhood RED solution. *Springer
             Wireless Networks*, 11(4):383–399, July 2005.

[XS01]   Shugong Xu and Tarek Saadawi. Does the IEEE 802.11 MAC protocol
             work well in multihop wireless ad hoc networks? *IEEE Communications
             Magazine*, 39(6):130–137, June 2001.

[YKT04]   Zhenqiang Ye, Srikanth Krishnamurthy, and Satish Tripathi. Use of
             congestion-aware routing to spatially separate TCP connections in wire-
             less ad hoc networks. In *MASS '04: Proceedings of the 1st International
             Conference on Mobile Ad hoc and Sensor Systems*, pages 389–397, October
             2004.

[YLN03]   Jungkeun Yoon, Mingyan Liu, and Brian Noble. Random waypoint con-
             sidered harmful. In *INFOCOM '03: Proceedings of the 22nd Annual
             Joint Conference of the IEEE Computer and Communications Societies*,
             volume 2, pages 1312–1321, March 2003.

[YS07]   Yung Yi and Sanjay Shakkottai. Hop-by-hop congestion control over a
             wireless multi-hop network. *IEEE/ACM Transactions on Networking*,
             15:133–144, February 2007.

[YSY03]   Luqing Yang, Winston K. G. Seah, and Qinghe Yin. Improving fairness
             among TCP flows crossing wireless ad hoc and wired networks. In *Mobi-
             Hoc '03: Proceedings of the 4th ACM International Symposium on Mobile
             Ad Hoc Networking and Computing*, pages 57–63, June 2003.

[Yu04]   Xin Yu. Improving TCP performance over mobile ad hoc networks by ex-
             ploiting cross-layer information awareness. In *MobiCom '04: Proceedings
             of the 10th Annual ACM International Conference on Mobile Computing
             and Networking*, pages 231–244, September 2004.

[YYS⁺04]   Taichi Yuki, Takayuki Yamamoto, Masashi Sugano, Masayuki Murata,
             Hideo Miyahara, and Takaaki Hatauchi. Improvement of TCP throughput
             by combination of data and ACK packets in ad hoc networks. *IEICE
             Transactions on Communications*, 87(9):2493–2499, September 2004.

[Zad65]   Lofti A. Zadeh. Fuzzy sets. *Information and Control*, 8:338–353, 1965.

[Zad68]   Lofti A. Zadeh. Fuzzy algorithms. *Information and Control*, 12:94–102,
             1968.

[ZCF05]   Hongqiang Zhai, Xiang Chen, and Yuguang Fang. Rate-based transport control for mobile ad hoc networks. In *WCNC '05: Proceedings of the IEEE Wireless Communications and Networking Conference*, volume 4, pages 2264–2269, March 2005.

[ZF06]   Hongquiang Zhai and Yuguang Fang. Distributed flow control and medium access in multihop ad hoc networks. *IEEE Transactions on Mobile Computing*, 5(11):1503–1514, November 2006.

[ZLX02]   Li Zhang, Zhen Liu, and Cathy Honghui Xia. Clock synchronization algorithms for network measurements. In *INFOCOM '02: Proceedings of the 19th Annual Joint Conference of the IEEE Computer and Communications Societies*, pages 160–169, June 2002.

[ZSZ03]   JianXin Zhou, BingXin Shi, and Ling Zou. Improve TCP performance in ad hoc network by TCP-RC. In *PIMRC '03: Proceedings of the 14th IEEE International Symposium on Personal, Indoor and Mobile Radio Communications*, volume 1, pages 216–220, September 2003.

Index

www.ingramcontent.com/pod-product-compliance
Lightning Source LLC
LaVergne TN
LVHW062313060326
832902LV00013B/2185

* 9 7 8 3 8 3 6 4 5 2 7 9 3 *